The Invisible **Power** within Foods

A Comparison of ORGANIC & NONORGANIC

Includes insight into genetically modified foods

Crystallisation images from research by LifevisionLab of Soyana

A. W. Dänzer

IMPRESSUM

A. W. Dänzer

The Invisible Power within Foods

A Comparison of Organic & Nonorganic

Includes insight into genetically modified foods

Crystallisation images from research by LifevisionLab of Soyana

Foreword by Ruediger Dahlke, M.D.

First English printing in October 2016

Translated from German and edited by Hidde Tangermann and Eric Mullerbeck

Published originally in German in October 2014 by Verlag Bewusstes Dasein, Schlieren-Zürich, Switzerland

Author and project leader: A. W. Dänzer

Layout, typography and setting: Soyana, Zürich

Printed in the Czech Republic on environment-friendly, chlorine-free bleached paper.

All photo material of microscopy images incl. cover photos: LifevisionLab of Soyana, Schlieren-Zürich, Switzerland

Photo credits: Adarini Inkei [9,11], Reinhard Geßl [15], Otto Hermann [15], Stoyko Sabotanov [17], Giorgio Mercandelli [60], Lido Vannucchi [62], Coop Switzerland [249], Phyllis Hart [251], Whitmuir Farm [255], Dreamstime Stock Photography and public resources from the internet.

ISBN 978-3-905158-17-5

Verlag Bewusstes Dasein Walter Dänzer, Turmstr. 6, 8952 Schlieren-Zürich, Switzerland

Tel. +41 (0)44 731 12 00, Fax +41 (0)44 731 12 75, Email: info@verlag-bewusstes-dasein.ch, www.verlag-bewusstes-dasein.ch

Visit the website for this book...

in English: **www.organic-nonorganic.info** / in German: **www.bio-nichtbio.info** / in Italian: **www.bio-nonbio.info**

Contents

Foreword

> *Information and coherence in our food are the keys to a long and healthy life.*

I am pleased to provide a preface to a book on nutrition that starts with a philosophy of life. In my work I have often seen that traumatic experiences can be the basis for disease. The idea that disease patterns simply reflect, or indicate the presence of, deeper psychological or spiritual conflicts is founded on this insight, and in the spirit of true psychosomatic medicine puts the soul before the body. Nonetheless, psychosomatic medicine also attaches great importance to the body and its nutrition using plant-based and wholesome "peace food".

Medical studies taught me very little about nutrition, most of it being either false or shortsighted. Conventional medicine only recognizes the physical aspect of food, which amounts to merely counting calories. It cannot grasp the concept of quality and is unable to distinguish organic from conventional food. Even the vegan scene often has problems there. White flour and white sugar may be vegan, but they are not healthy.

If we do not grasp the concept of food quality, we cannot understand the importance of freshness, yet both are essential for a healthy diet. This became clear to me when I was working on *The Secret of Life-Energy in our Food*. Information and order in our food—both closely connected to the sunlight stored in food—are the keys to a long and healthy life. Both can be tasted, yet their existence cannot be proven scientifically.

This book represents a decisive step forward by making food quality visible, and revealing the order in food in a meaningful way. Organic food distinguishes itself clearly from the chaos found in conventional, industrial food. Erwin Schrödinger, winner of a Nobel Prize in physics, considered the sunlight stored in food as the key factor of its order.

The crystallization technique presented here reveals the inner order of organic food through beautiful images that speak for themselves. According to Nobel Laureate Ilya Prigogine, we rely on constantly absorbing order, in order to live and grow.

Through different methods, such as copper-crystallization tests, anthroposophical researchers have long tried to define quality. The method used by Swiss company Soyana's laboratory leans on old spagyric procedures*) and succeeds in making food quality visible for everyone. Even just an initial glance at the results will reveal the sky-vast differences between conventional and organic vegetables. Even small children can immediately recognize the beauty of the images of organic plants. The method presented here is standardized and free from arbitrariness, representing a quantum leap beyond the ice-crystal images of Emoto. It convinces through its simplicity and the appealing beauty of the photographs. For the first time organic food quality has become clearly visible.

— Ruediger Dahlke, M.D.
Doctor, psychotherapist and author

*) Spagyric procedures refer to a method in which the ash of a medicinal herb is added to the extract of the herb in order to enhance its healing properties.

The Story

How this book was born

This book is the result of a thirty-year journey. After our company Soyana created the first Swiss meat substitutes in 1981, I started asking myself the question: What is the true value of food? Over the years the answers to that question have ripened through my daily practice of meditation and were illumined by inner and higher gifts: food is consciousness, and purity and love are the most important qualities for those working with it. The water research by Masaru Emoto in the 1990s made us aware of how important the energetic quality and information-conveying coherent structure/order of water are for our foods.

In order to learn more about these qualities, as well as to improve Soyana foods, we founded our own laboratory for subtle food quality research in 1996. Little by little we discovered not only how to energize the water for our foods, but also how to create harmony in the food-production environment, including the building, electricity and gas, and even the air, so that we could truly ensure the holistically good quality of the Soyana foods.

In the year 2000 we made the film *Über-Lebens-Mittel (About Life in Food)* for the 20th anniversary of Soyana. The 40-minute film introduces the four most important Soyana Quality Principles:

1) plant-based foods, 2) from organic agriculture, 3) possessing good life force or subtle energy and 4) manufactured in a good consciousness.

In 2001 the Soyana team showed the film to our customers in a series of 100 evening programs in all parts of Switzerland, with great success. Every evening many customers congratulated me with a heartfelt handshake and begged me to please continue our work exactly along these lines. (The film can be seen in German, French and Italian on *www.soyana.ch* under the heading *Qualität*.)

Thousands of LifevisionLab pictures have shown us that there are striking demonstrable differences between different foods. (See page 20 for more information on LifevisionLab.) We discovered both wonderful and horribly ugly images, and I began to notice correlations with the cultivation and processing methods of food. The idea was born to make these pictures available to consumers on food labels, so they could make better informed decisions with regards to their food purchases.

I submitted this idea to the Federal Office of Public Health (in German *Bundesamt für Gesundheit, BAG*) in Switzerland, which is responsible for food labelling standards. I asked them to initiate a working group to study the details of how the life design principle*) could be shown on the labels. I presented them with a considerable amount of documentation. The woman I spoke with was personally very interested, but the BAG dismissed the initiative as "the method was not

*) *Life design principle*: the fundamental organizing principle of life; the blueprint of the life-energy or life force in foods

acknowledged by mainstream science". When I asked her what exactly was meant by that, she answered that the BAG could only work with something taught by professors at the Swiss Federal Institute of Technology. I was disappointed, since that would probably never happen in my lifetime.

The rejection by the BAG experts, who were not even willing to talk about improving consumer information on food quality, puzzled me. In their official statements they consider consumer deception to be the worst crime, and rightly so. But while deception is still happening every day in millions of cases, the BAG does not offer any means through which the consumers can avoid this deception. From this experience, it seemed to me that the experts and government officials in the field of food quality do not have a sincere interest in the subject, but that their main concern is to maintain a stable status quo, in order to secure their own position. Clearly, the only way for me to serve the cause of improving food quality is to work directly for and with the consumers. Only consumers are sincerely interested and motivated. It is the consumers who have to be informed, since through their purchases they ultimately decide which foods are being produced. Again, a well-known truth stared me in the face: only the pressure of consumers is able to bring about change. From the point of view of the status quo this change will be called a revolution, but in reality it is only the implementation of a necessary improvement whose time has come.

For me the way ahead was clear: I had to write a book in order to reach as many consumers as possible. It was also clear that I had to finance the research myself, and that it had to be executed by our team, even though projects of this scope are usually carried out by institutions who receive large subsidies. On the other hand I am grateful that we have succeeded in finishing the project, even though it required much free time and patience, since other work often took priority. It is gratifying that groundbreaking research can still be achieved by single individuals.

It is my intention to open up a window into a domain of reality that until now has been unrecognized. The view through this window will trigger many new insights and generate thousands of new questions, which naturally will have to be studied and answered. Further suggestions and resources regarding these matters can be found at the end of the book.

It is a great joy to see these pictures inspire more and more people by showing them the amazing beauty of creation, and to see the awakening realization that we can nourish ourselves with the rich beauty and the natural health qualities in food, if we act intelligently.

– A. Walter Dänzer
Founder and leader of the family enterprise
Soyana and of the LifevisionLab

Health and peace

Health is a state of complete physical, mental and social wellbeing.

– Definition of personal health by the World Health Organization of the UN (WHO), contained in the WHO constitution: *www.who.int/governance/eb/who_constitution_en.pdf*, accessed April 2016.

Health and peace are our most basic needs. When we endeavor to achieve them for all of humanity, we will receive what we personally long for most: happiness.

WHO's emphasis on positive wellbeing instead of the absence of disease testifies to the profound insight and great wisdom of the authors. It implies important truths about the essence of health:

- **Health is an experience, an experience of wellbeing that can be increased in infinite measure.**
- **Wellbeing creates satisfaction, an experience of fulfilment that radiates peace and happiness. Wellbeing is followed by positive and harmonious thoughts and deeds.**
- **The deeper understanding of health is a key to a fulfilling life, which brings peace and harmony into the world.**

We admire the wisdom of the authors and of all those persons that voted for this definition at the UN Forum of 1948. May we follow their wisdom and help to achieve its ideals.

My happiness is my healthiest achievement.

—**Sri Chinmoy**, Meditations at the United Nations, 2005

Five practices for wellbeing

Maintaining my personal wellbeing through the practice of a holistic lifestyle, which is a natural protection against disease.

On the previous page we saw that health is clearly a state of wellbeing that is closely connected to peace. Again, peace is not only a global goal, but also a personal path, part of a holistic lifestyle that offers us peace and harmony with ourselves and the earth.

We can take good care of our health by consciously and daily becoming active in five different areas.

Plenty of movement. Plenty of physical movement in the sun and the fresh air. In our modern society with its predominantly sedentary lifestyle the health benefits of physical exercise have been generally and widely recognized, but are still insufficiently practiced.

Healthy food, body care and hygiene. Free access to fresh, healthy food, pure water and adequate knowledge and facilities for body care and hygiene. These are often postulated human rights, which for a large portion of humanity are sadly still a far cry.

Good life-energy. Good life-energy in all sorts of living spaces. Very few people are aware of how much a living environment full of free-flowing, fresh life-energy contributes to our health. Our wellbeing is compromised by electrosmog, objects that block the life-energy and materials that are hostile to life as well as naturally occurring 'zones of disruption' that interfere with our life-energy. Most people are hardly aware of these influences. But a pure subtle atmosphere full of flowing life-energy is just as much a basic human requirement as access to clean drinking water. Companies like Soyana with the Ojas Harmony System offer solutions to these problems, which have brought much welcome relief to many people. Yet awareness is still generally lacking about the importance of creating public spaces—schools for instance—that contain healthy life-energy.

Positive contributions in the social sphere. Offering positive contributions and fostering harmonious relationships within the family, in social groups and with humanity in general. To do something positive for other people every day contributes significantly to our own personal wellbeing and is not a charitable action for which we should expect special gratitude. Social wellbeing implies that we feel part and parcel of the human family, and we act correspondingly as a part of the whole, which far transcends our littleness.

Practicing an inner life.

Discovering within ourselves our spiritual heart, where we feel one with ourselves, the Creator and the creation. When we learn to go deep within our heart, we can receive infinite light, peace and joy from the source of life. In deep meditation we experience ourselves as limitlessly happy beings in a sea of infinite consciousness.

This higher world is not accessible from the unillumined consciousness of the body, vital and mind. (The term "vital" is used by Sri Aurobindo and Sri Chinmoy to refer to the portion of our own being that embodies the universal life-energy and personal life force and commonly expresses itself through emotions. The root of the word "vital" is *vita*, which means life in Latin.) In the course of our self-improvement, we as seekers learn about the many levels of consciousness in our own integral reality, and we also learn how we can act with self-giving love to fulfill our role on earth.

Through the daily practice of the inner life, we develop our full potential. To shape this lifelong inner journey satisfactorily and successfully, the seeker needs an inner hunger and a qualified spiritual teacher.

Healthy food

From among the five practices for wellbeing, in this book we focus on healthy food. The focus is not on various diets, but rather solely on the quality of the foods that nurture our health.

Whatever diet we choose, we always require healthy water and healthy food.

When we ask ourselves what is the value of food apart from the material nutrients (like protein, fat, carbohydrates, vitamins, minerals, phytonutrients, etc.) it provides, then the question arises: is there an invisible power, innate in all foodstuffs, that keeps us naturally healthy? As humans we are not merely an assembly of raw materials, but a highly integrated living organism, and this very same life, which is invisible, clearly belongs to a dimension of reality beyond the purely material. The question is: are foods a mere assembly of matter or is there something more, such as an invisible life-energy, and a coherent, ordered template conveying essential information, which is a fundamental part of every living thing?

» What is healthy water? What are healthy foods? What is the true value of water and food?
It is their life-giving, life-supporting capacity.

In order to reveal the life-giving and life-supporting capacity of water and food, and thus make it observable, Soyana founded its own laboratory LifevisionLab in 1996 for research into higher food quality. The tasks of Soyana's LifevisionLab were:

- **To make visible the subtle life-energy or order force of food;**
- **To educate the Soyana team about life-energy in water and food;**
- **To enable the Soyana team to create Soyana foods with pure life-energy and enhance plus examine their quality.**

At first the purpose of the LifevisionLab was only to provide internal support for Soyana's work. Yet when thousands of photographs (taken according to the processes outlined in this book) demonstrated clear differences between organic and nonorganic foods, it became clear that these findings should be made available to a larger public. Thousands of farmers and gardeners and millions of consumers would be interested in them. We are therefore undertaking to publish the results of our research, so that consumers will be able to make informed decisions in their food choices, and so essential criteria for judging food quality, which up until now have remained invisible, can be made visible and openly considered.

In this way the idea for this book was born.

What is ORGANIC? What is NONORGANIC? *)

When farmers do their job well, inquisitive consumers are not always able to tell for certain, by examining the food, whether it is organic or nonorganic, although sometimes there are clear differences. Even analyzing the nutritional value sometimes fails to offer the necessary clarity, although generally organic foods have better nutritional values and a purer taste.

But in almost all cases, using the techniques developed at Soyana to look at the life design force or order force—the inner life-energy that guides the growth of each living entity—will make things immediately clear for any particular food.

» **Outer appearances may deceive, but not the inner information.**

One of the most important "employees" of organic farmers in cultivating soil fertility is the earthworm. Not only does it create tunnels for air and water, but also it refines and enriches the life present in the soil. On the right its typical structure, the earthworm casting made of fine clay, humus and its metabolites.

Photo left from "Humus - regulator full of secrets": Reinhard Gessl/*www.biola.at*
Photo right: Otto Hermann/Bildarchiv Boden

ORGANIC...

...means that the plants grow in a holistically cultivated and life-friendly soil in which they feel well and thrive.

In organic agriculture human action focuses on the all-encompassing support of life, in order to ensure that plants develop well.

Organic cultivation means that the life in the humus and good environmental conditions for the plants are strengthened. Use of compost, rock dust, ground cover consisting of organic materials, and mixed cropping all benefit the life in the humus, which includes micro-organisms, earthworms, and thousands of other life forms.

The organic mindset embraces and works with the whole natural cycle, with consideration for all parts of the whole, which is why it can be called wise: the powerful life-energy in the entire natural cycle, especially in the humus, boosts the plants and gives them strength.

The results are manifold: healthy plants have fewer weaknesses, giving parasites fewer chances against them. Therefore, there are much fewer parasites. Chemical agents can be avoided. The natural and uncontaminated foundation of the plants remains intact and is constantly renewed. Organic agriculture is sustainable.

This is also made clear by a quote from a publication by the Research Institute of Organic Agriculture (FIBL) in Switzerland:

Organic means concern for a living soil

The essence of organic agriculture is a living soil, which means the soil is rich with flora and fauna, helping to stabilize the surrounding ecosystem. Contrary to intensive agriculture, which partly replaces the natural cycle with artificial fertilizers and pesticides, organic agriculture tries to work, not against, but rather in harmony with nature. It combines a series of methods that directly benefit building up more humus by using natural substances to support organic life-cycles, together with various crop rotations, permanent soil cover, organic compost and reduced soil cultivation.

In organic agriculture living soil is also sustained by mixing different crops. For example, the different root systems of legumes and camelina (an oilseed crop) complement each other. Above the ground these plants form a protective lattice, which is resistant to wind and weather. As a result, plant diseases can be better prevented. Spraying pesticides and intensive soil cultivation become unnecessary.

*) Information about what is ORGANIC and what is NONORGANIC is offered by all organic organizations on their websites and in their publications.

NONORGANIC...

...means that the crops are primarily nourished with dissolved chemical fertilizers, whereas everything else that lives in the same space (weeds, fungi, molds, insects, parasites) is considered a competitor and is treated with herbicides, pesticides, fungicides, etc. to destroy it.

In nonorganic agriculture human action focuses directly on the results that humans want to obtain from plants, i.e. the greatest possible yields, attractive appearance, long shelf life, and specific features such as high protein or fat content.

Nonorganic agriculture works with the framework of a scientifically nourished plant and analytically defined approved thresholds, based on toxicological studies, for the residues caused by chemical additives. It typically requires large amounts of raw materials and machine energy, and a continuously declining amount of labour. Chemical agents are used to 'wage war' and destroy the competitors of agricultural crops. The battle is pursued with extreme measures, similar to a war with no concern for collateral damage.

The use of chemical fertilizers started with the chemist Justus von Liebig, who discovered that plants grow faster and their yields increase when they are fed with chemical nutrient solutions. Nonorganic agriculture is based on a scientific, rational and functional approach that fails to notice many aspects of the entire reality, which it is therefore also unable to shape.

This primarily short-term and economically motivated approach has produced tremendous economic successes. It has transformed agriculture into an agro-industry that produces our food with the aid of a high consumption of energy, chemical substances and machines, and a decrease in human labour, and does so very efficiently and cost-effectively—as long as one overlooks the damage done to the soil, to groundwater and other bodies of water, to biodiversity, and to the health of humanity. It does not apply the 'polluter pays' principle. The damage has to be paid by the general public and by those affected by the specific consequences listed above. Never before have we been able to grow so much food with so little labour. And never before have we been able to feed such a large human population.

Yet despite the quantitative sufficiency of our food, there is a steady qualitative decline of minerals, vitamins and micronutrients in nonorganic food which can lead to a deficiency of essential nutrients in the nutrition of human beings, making the qualitative aspects of food ever more important.

Over time the agricultural use of chemical agents and heavy machinery has had serious consequences:

- **The decline of soil life and the poisoning of soils, groundwater, rivers and even oceans.**
- **The problem of residual poisons in food, resulting in the contamination of the entire food chain.**
- **The toxic overload in the organs of biochemical processing and elimination, both in animals and humans.**
- **The pervasive and often preventive use of antibiotics in industrial livestock farming, leading to a potential public health emergency through the rise of uncontrollable and dangerous antibiotic-resistant bacteria.**
- **The severe and fast-advancing worldwide loss of fertile soil.**
- **Conventional industrial agriculture's heavy dependence on fossil fuels. This makes it clearly non-sustainable, presenting huge health, ecological and even existential problems for the future of humanity.**

In addition to the above serious consequences, this book introduces evidence for an additional major concern:

> *The weak and often largely extinguished life design principle (also called life-energy or order force) of nonorganic foods.*

This book offers insights into the existence of life-energy in food, and it also draws the attention of the general public to the fact that life-energy in nonorganic foods is strongly decreased.

From now on, every serious assessment of food quality will have to consider these new criteria. New possibilities for shaping our own destiny are opening up to us. How do we want to deal with the reality we have just become aware of?

Of the 1,150 chemical-synthetic agents currently used worldwide in nonorganic agriculture, 451 are considered especially dangerous.
Source: *www.bio-wissen.org*

Summary

- **The nonorganic mindset in so-called conventional agriculture is focused specifically on short-term economic success with little consideration for essential components of the whole natural cycle, which is why it can be called one-sided.**
- **The high quantity yields of food products in the West come at a heavy cost, causing tremendous damage to our quality of life and the ecology of the entire earth.**
- **One of the disadvantages of nonorganic foods—and one not until now taken into account in debates over organic versus nonorganic—is their weakened life-energy or order force.**
- **The weakened life-energy of these foods fails to sustain our wellbeing as we face the many harmful influences of our modern-day lifestyle, so that we are increasingly incapable of meeting these challenges.**

The work presented in this book offers a first view of a completely new approach to seeing the hitherto invisible power in food, so that the observers of these images may form their own judgements and draw their own conclusions by answering the following questions.

The Questions

Are organic and nonorganic foods very similar or significantly different in their life-energy or life design principle?

If a difference can be observed, what is it exactly? How can it be described?

Does it follow from the observations that organic foods have more clarity, beauty and order in the expression of their life-energy or life design principle, and that nonorganic foods are weakened in this respect when compared to organic foods?

Fundamental correlations in nature:

Function, order, beauty and health

When we observe the structure and physiology of a plant, we are amazed at how efficiently and yet simply a plant supplies each of its parts with water and nutrients and carries away metabolic waste products. When we compare plants to man-made structures such as a city, we immediately notice that in the plant, all the transport happens on the inside: from the root to the smallest leaf, both large and finely branched channels lead the essential substances to where they are needed, supplying all cells of the entire organism in a perfect way. We can see this quite clearly in a leaf or in an entire tree.

Nature thus designs its structure on the basis of its needed function. An ordered system arises, that ensures all the life fluids can flow and that nothing is blocked or wasted. In this way the plant can unfold all its physical processes very efficiently, from growth to blossoming to bearing fruit. Thus it achieves optimum health and a natural resistance against attacks. We humans find this order beautiful. In nature beauty thus arises out of proper functioning, which in turn arises out of an order that shapes the total plant body and ensures the health of the organism.

In the world of plants, beauty and health are the results of a functional order; from the very outset these qualities are rooted in an intelligent design that performs the function it is designed for. As long as the organism is able to maintain its order, beauty and proper functioning, the health of the organism is also assured.

Since the plant and animal kingdoms share a common cellular ancestry, we can observe the same principle in our own organism, for instance in our internal supply system, which is the cardiovascular system.

We consider this naturally developed order to be beautiful. Why? Is it because it is deeply imprinted in our consciousness, and we recognize ourselves in nature? We feel so well in nature because the same natural order, which is actually a force of beauty, has created us as well. We are part of nature. The more we consciously recognize the many functions of leaves, trees and forests, from the use of sunlight to the production of chlorophyll, transmutation of carbon dioxide into oxygen, maintenance of water balance, creation of habitat for millions of other species, climate regulation via a natural 'one hundred percent recycling policy' of all organic materials, the more we are amazed at how intelligently nature has organized itself. We admire the expression of an all-dimensional design immediately as the creative power of a higher intelligence.[*)]

*) Even when many scientists argue with Darwin and cannot perceive a higher intelligence at work, people seeking for answers now find support in an entirely new scientific field. For more information watch the highly interesting documentary *Unlocking the Mystery of Life: The Scientific Case for Intelligent Design.*

You can see the film at *www.lifevisionlab.org* (click on *Links*)

(Please be assured that I have no intention of entering the battleground between Darwinists and creationists. However, as a practitioner of yoga and meditation, I experience life as an ongoing creation by a conscious Creator within, which is completely real and alive. This film confirms my personal experience. Therefore, I find it remarkable.)

1

Experiment One

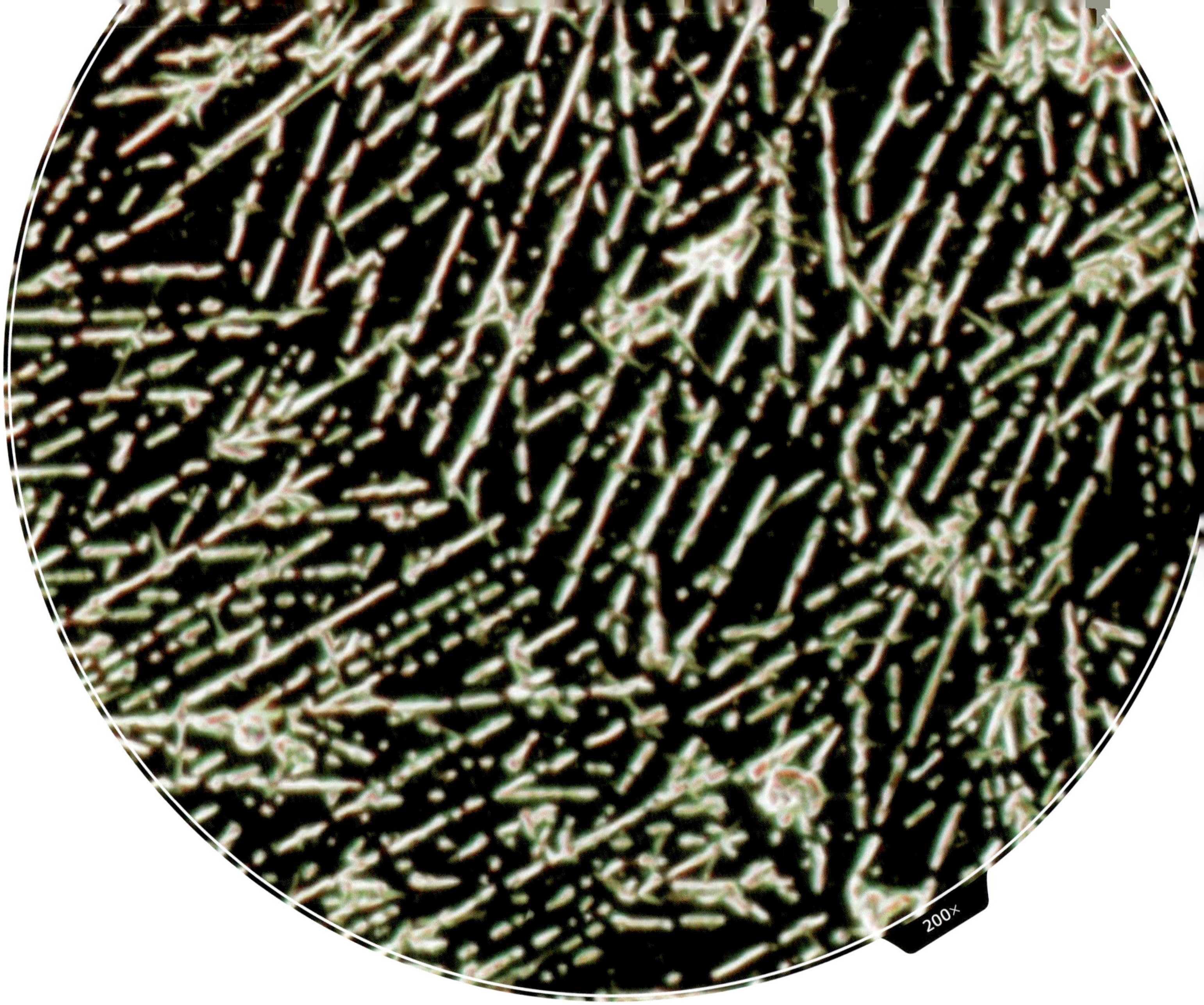

Which green tea is organic?

Compare these two microscopic photographs. What you see are two images of the life-energy in green tea, one organic, the other nonorganic. (The organic green tea is used as an ingredient in Soyana's fermented Green Tea Chi drink.) These pictures were shown to hundreds of consumers and dozens of children, and the following question was posed...

Which green tea is organic?
The harmonious one with the plant-like twigs.

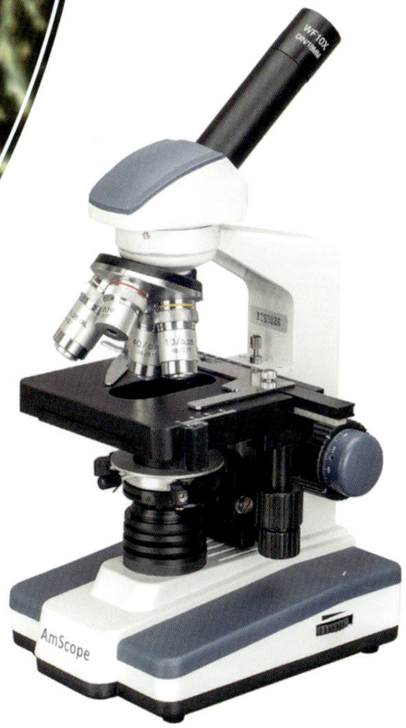

Three important observations and one consequence:

1 This survey did not identify even a single person unable to answer the above question correctly, including children. Is it not phenomenal?

2 There are great differences between organic and nonorganic foods: in the latter the energy doesn't flow naturally; the lovely round shapes have become square and rigid. In nonorganic foods something has changed. The organic here becomes nonorganic: mineral, like the fertilizer that is used in artificial agriculture. The consciousness of the plants becomes mineral or stone-like. Does this benefit humans and animals? What are the effects?

3 The human capacity to recognize visual structures is so amazingly well developed that no education is required. Each child is automatically and intuitively familiar with the process. No training is needed for us to interpret the LifevisionLab photographs of the life-energy within foods correctly.

Consequence: Imagine that the life-energy of a food were shown on its food label. Would we pay attention to it, so we can make informed decisions about our food purchases?

2
Experiment Two

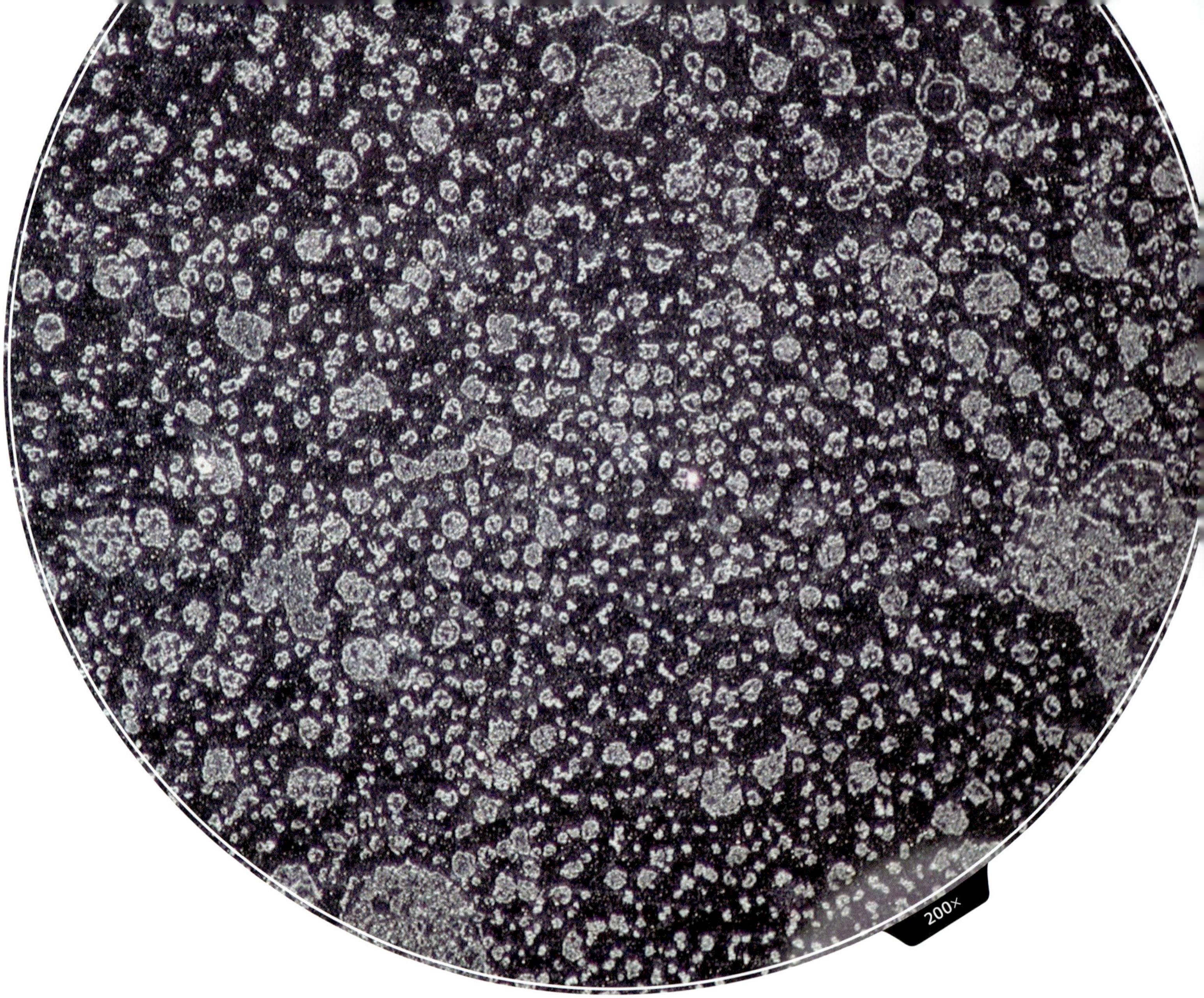

Which drink is organic?

Here are two pictures of the life-energy in popular drinks. One is organic, the other is not. Both are processed foods. These pictures were shown to hundreds of consumers and dozens of children, and exactly the same question was posed...

Which drink is organic?
The beautiful, flowerlike one.

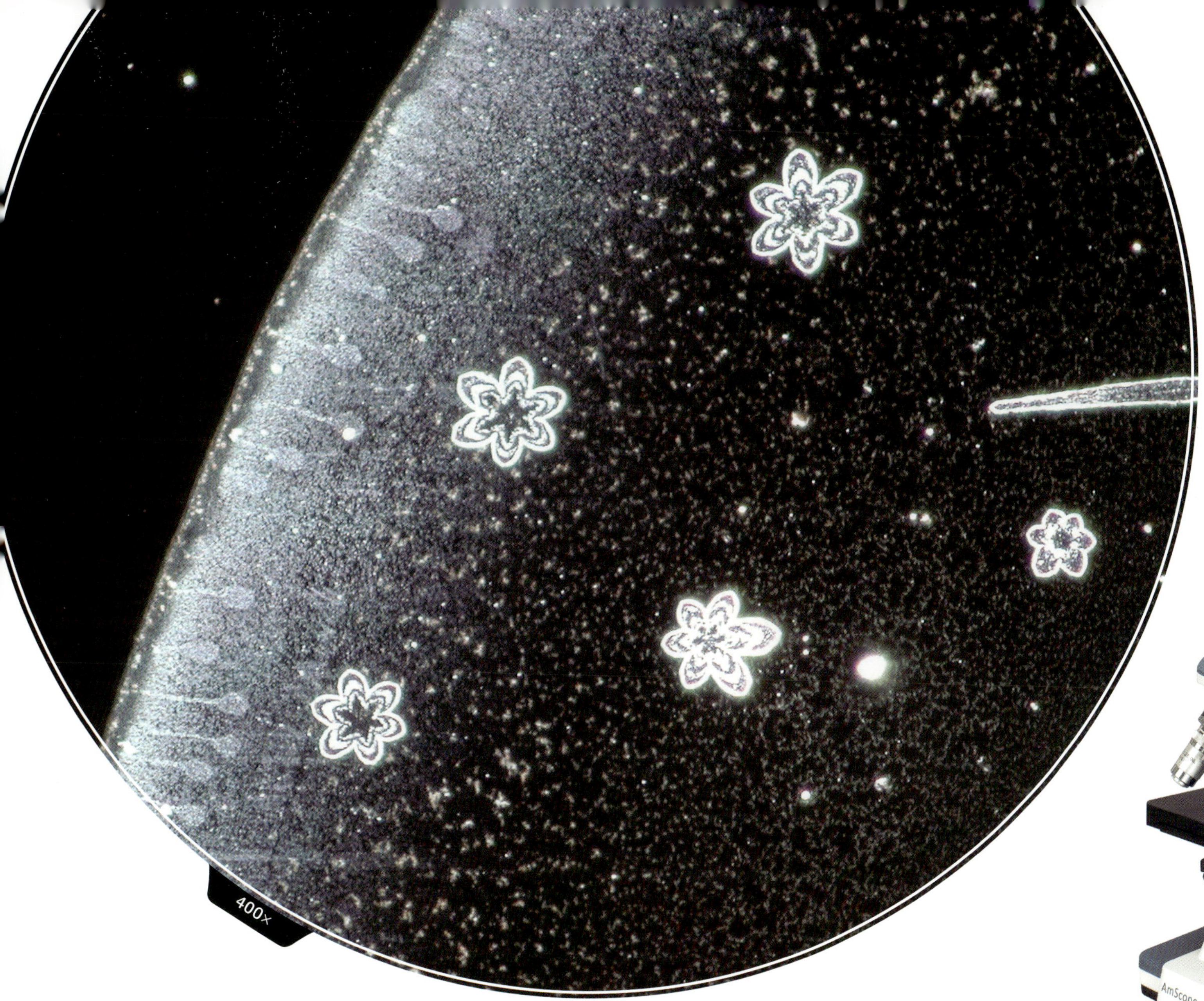

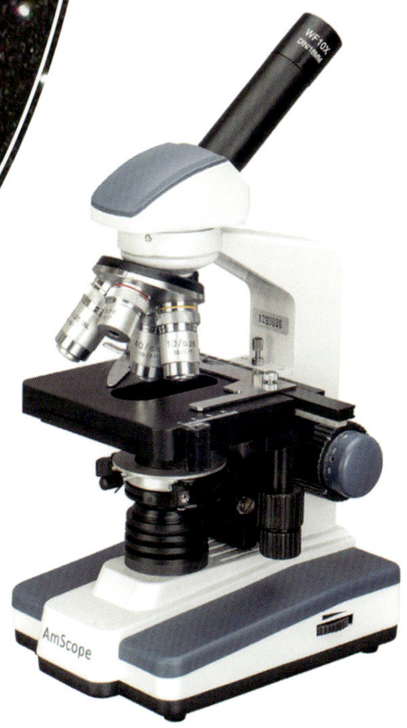

Three astonishing observations and one important lesson:

1 Once again the survey did not identify a single person unable to answer the question correctly, including children. For human beings this is self-evident. We quickly grasp the logic.

2 Even in processed foods there are great differences between organic and nonorganic ones: the structural arrangement of the nonorganic drink is entirely amorphous, unorganized and without any sign of life. It looks like a pile of debris. Conversely, the organic drink reveals six-petalled multilayered blossoms that beautifully unfold. It will interest the reader here to know that in alchemy six-petalled blossoms are called "flowers-of-life", as far back as the Middle Ages. How did they know? (Alchemy was "all-chemistry", in which the seekers of the secrets of life were holistically oriented.)

3 Again the human capacity to recognize visual structures is shown to be so amazingly well developed that no education is required. Each child is automatically and intuitively familiar with the process: everyone interprets the LifevisionLab photographs of the life-energy within foods correctly.

Lesson: The visual language of the LifevisionLab is UNIVERSAL and its understanding is innate. These pictures are thus well suited to performing important tasks, such as characterizing and labeling foods in order to communicate their quality.

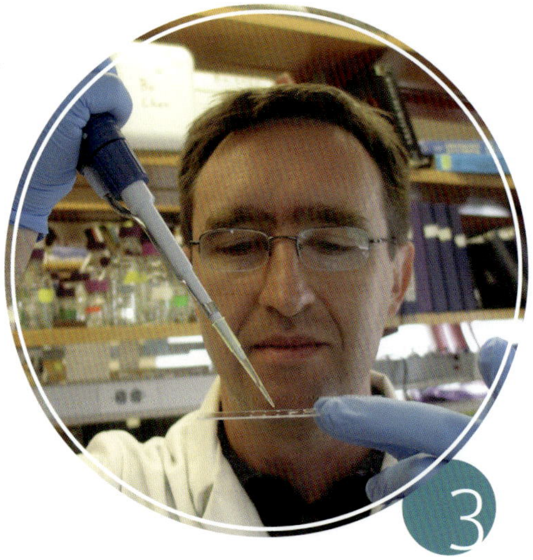

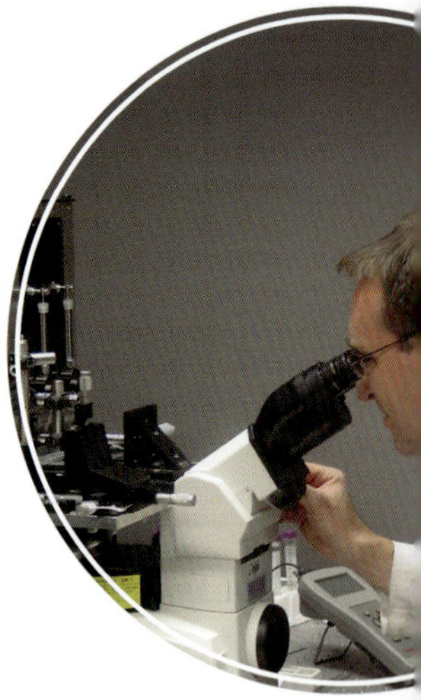

The Soyana Method

Crystal images revealing order

The Soyana Method, perfected and standardized since 1996 by Swiss food pioneer Soyana, is the method of researching the life-energy or order force (life design principle) in water, food and other substances.

Before looking for the most suitable method for us, we already knew the wonderful results published by our great predecessors. Among those are the image generating methods inspired by Rudolf Steiner, such as crystallizing drops of food on a metallic matrix of copper chloride, or Masaru Emoto's experiments with frozen water, as well as Ehrenfried Pfeiffer and other naturopaths who researched the properties of water and also bodily fluids for healing purposes. Although their results are wonderful and convincing, our idea was different: We were looking for a process whereby a suitable image would arise solely from the water or food substance itself, not from a metallic matrix that yields images that can only be interpreted by those with expert knowledge. The requested method was to create images that can be understood immediately by anyone. We hoped to find a method that would be able to investigate every kind of food and show infinitely different images according to the characteristics of the food. Ideally this method would allow one to know the essence of foods, not just one special feature in a single crystal, like Emoto's water crystals, but the whole bandwidth of features, an entire crystal landscape, so to speak.

In this way, we discovered the ancient spagyric method. We developed it further into a precisely standardized extraction process, until we were finally able to achieve the desired result through the so called SAT-Protocol (Soyana Advanced Technology Protocol). We simply call it the Soyana Method.

The Soyana Method is a process that studies the infinite behavior and potential of the clusters and sedimentations of minerals and other substances. Inspired by the originally European spagyric concept of energy and alchemy (holistic chemistry), and further aided by ideas derived from

Subtle-Energy-Science, ultra-high dilution physics and quantum physics, it has been further developed into a modern concept by Soyana's LifevisionLab research team.

In simplified form the process can be described as follows:

Using a precise standardized protocol, a test liquid is extracted from a particular food item. The purpose of the extraction is to distinctly bring to the fore the life-energy or order force of that particular food. (Duration: one day)

Droplets of this liquid extract are then placed onto a test tube, and dried and crystallized under specified, unchanging conditions. (Duration: one night)

The dried drops are studied and photographed under a microscope; the photos are then archived.

This method has several benefits. The photos allow us to observe entire crystal landscapes, which are nothing other than dried up minerals from the corresponding food item. But the life design principle or order force inside the food item determines how the minerals organize themselves. This allows for characteristic images for each food item, from among the infinite possibilities of how an image could form. At the same time these images can immediately be interpreted correctly and without any training by people who are open to new horizons, whether one wants to accept the method formally as scientific or not. For us the most important thing is to do a clean job, one that is replicable and that delivers messages that can be read and understood by all people.

For further study please refer to the expert contributions of Dr. Kausal C. Cortella, who for many years was the leader of Soyana's LifevisionLab. These are published on the website www.LifevisionLab.org.

A window opens...

In this book you will find microscope photographs of over 50 different foods, with a few exceptions both organic and nonorganic - a total of over 100 food items that have been prepared according to the aforementioned Soyana Method. In order to provide a wide perspective on the variety of order forces (life design principle) that are present in organic and nonorganic foods, we have chosen many common plant-based foods. Obviously the number of foods that could be included was limited. For reasons of simplicity we have focused only on plant-based foods.

Our goal is to open up a window into a heretofore unknown aspect of food quality. But we are not trying to present a general overview of all possible outcomes and differences among all types of foods. That is why mushrooms, sea vegetables, meat, fish and dairy products are not included in the research described here. Similarly we do not at this time consider the distinctions between the different special cultivation methods, such as organic vs. biodynamic, organic vs. natural Japanese cultivation and many more. These differences may be studied in future research. Hundreds of questions will arise after the publication of this book which we cannot yet answer, such as quality differences arising from the influence of the sun and the weather, a systematic research into the impacts during storage, manufacturing and transportation, etc.

Only three special cases were researched as single examples: foods growing in the wild, the influence of the microwave on a herb and the effect of genetic modification on the life force of soy beans.

A universal language

Although this book opens up a new perspective and reveals deep and essential insights, it can only be a small beginning. Subsequent research projects will have to clarify the hundreds of interesting questions that follow.

However, we are glad to present this research: both a humble beginning and a great discovery, that with the Soyana Method a new universal language for the evaluation of an important part of food quality has come into existence. For is it not fascinating that sometime in the future, with just one glance at such LifevisionLab images, anyone will immediately be able to tell whether a certain food appeals to them or not?

Thus we accomplish two important things. First, this research provides a method to expose the fraud of beautifully packaged foods with no life-sustaining energy within them. Today a widely used raw sales model, this trick will cease to be successful. The effect will be that food quality will improve, since the consumer naturally prefers good quality. And secondly, we consumers in the search for valuable information will no longer need to get a university degree in the study of food labels, since this method makes available to everyone a universal language that every child understands yet which can still convey highly complex content in a single image.

> *This comes down to enabling all people to choose the better option—a real revolution.*

With the following tips you quickly and easily learn the small alphabet of this new language, so that your journey through this book will be all the more enjoyable.

1 *Signs of life*

While looking at an image, you can simply identify yourself with it and feel what experience it gives you. Would you feel comfortable in that world?

Tap water from Queens, New York, USA

Tap water at Soyana in Schlieren, Switzerland, after being treated with the Ojas Harmony System

We have observed that all human beings have an innate capacity to interpret these images directly, and always correctly. This requires neither special knowledge nor training, since human capacities for recognizing and interpreting optical structures are by nature very highly developed. Life leaves behind traces of consciousness and beauty. We can all immediately recognize them.

2 Nature has taught us

All people recognize the creative power of life in great detail and with high accuracy. Since humanity has lived in nature's cradle for hundreds of thousands of years, our capacity to recognize and read the signs of life was schooled and shaped by nature and its creative power itself.

Life and beauty each one of us immediately recognizes in the life and beauty of nature. All human beings are themselves inundated with it and do not require any education in this field.

Fermented Chi drink

3 Where beauty speaks

We see life and beauty in an image when it looks like nature, when its proportions are graceful and its structure intricate. Such an image depicts a high degree of order force (life design principle). We experience ugliness or even impressions of hostility from an image in which the order force (life design principle) is lacking and the picture looks distorted, unnatural, degraded, disintegrating or even menacing.

Nonorganic cucumber

Japanese Nama Organic Miso (Soyana)

4 The order force depicted

Order force (life design principle) is the power that imbues the organism with order, coherence, symmetry and structure. It is weak when although the order is present, it is incapable of shaping the entire drop with its information; holes appear in the patterns depicted in the image. It is strong when the entire drop is shaped with its information, strong coherent structure and the impulses of the order force continue unabated, forming large clusters.

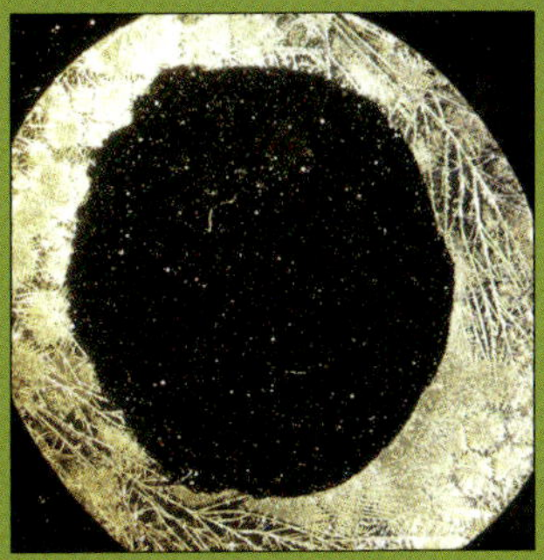

Nonorganic avocado

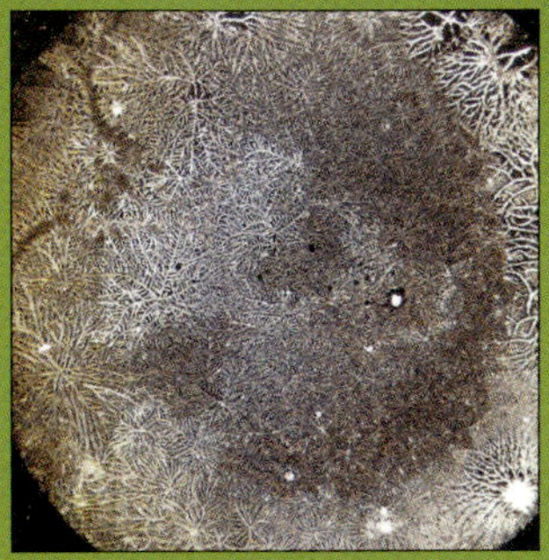

Organic avocado

5 Six-petalled blossoms

Six-petalled blossoms reveal a high degree of order force (life design principle). Already the ancient alchemists called them "flowers-of-life".

Fermented organic Chi drink

6 For your orientation

For good orientation this number shows the degree of microscopic enlargement: 200×

Life is evolving Beauty.
Truth is revealing Beauty.
Love is illumining Beauty.
Peace is manifesting Beauty.
Light is fulfilling Beauty.

—Sri Chinmoy

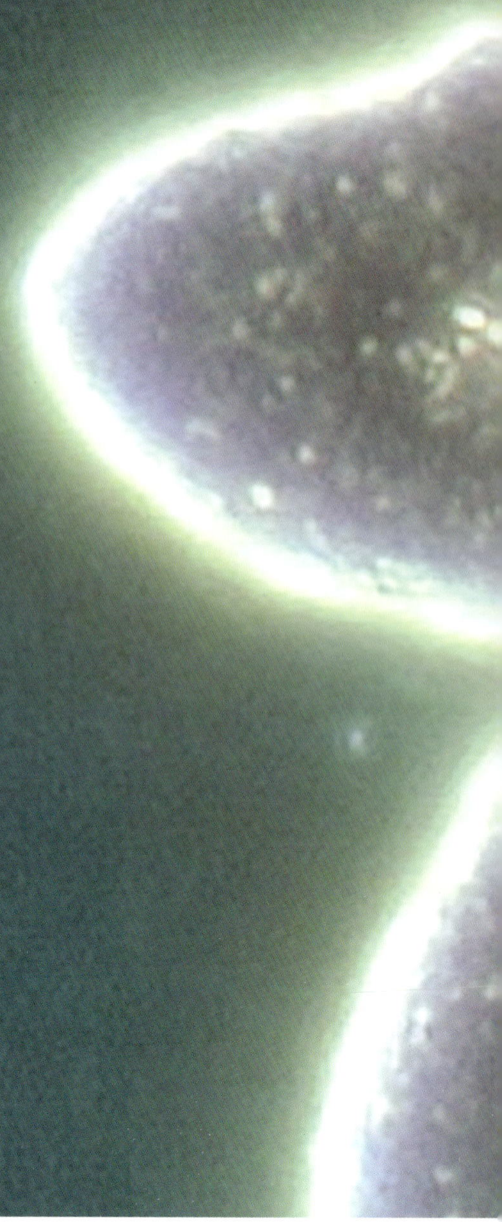

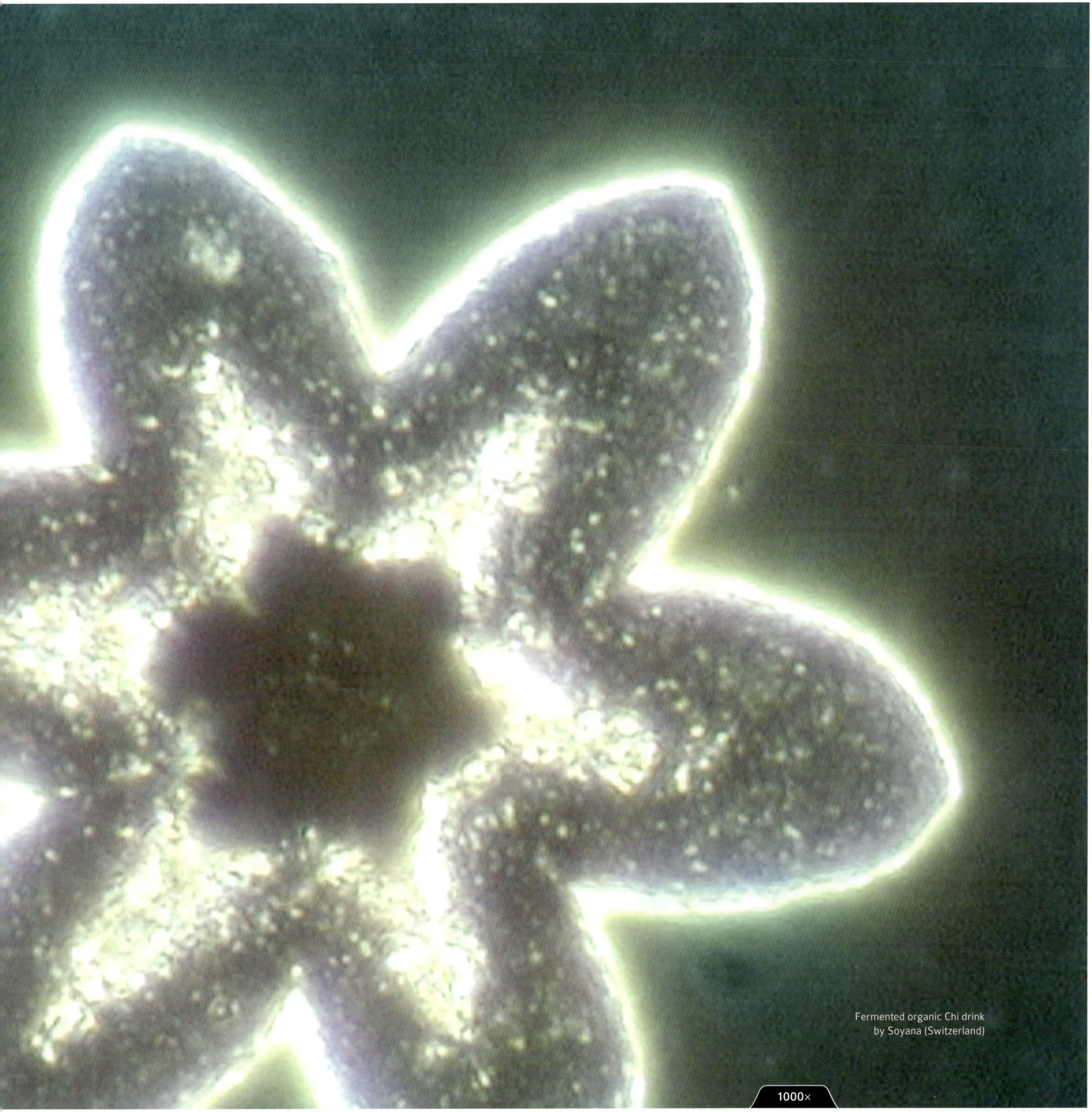

Fermented organic Chi drink
by Soyana (Switzerland)

1000×

Apple

Who would have guessed that within every part of an organic apple abides the figure of the entire apple tree, as well as the apple blossoms, seeds and even entire orchards of apple trees? The total natural life-cycle of the apple is holistically available and timelessly imprinted in every single drop of organic apple juice. It charms us with the wisdom of a fairytale, which tells us that the whole consciously breathes in every single part, and all the parts work together for the success of the whole, as if it was the most natural thing in the world. Our heart can understand and enter into this interdependent, all connected reality, whereas our mind struggles with it. We marvel and rejoice in the deep insight into the natural essence of things.

In the nonorganic apple this essential information is hardly visible anymore. It is blurred, lost and diffused into fragments.

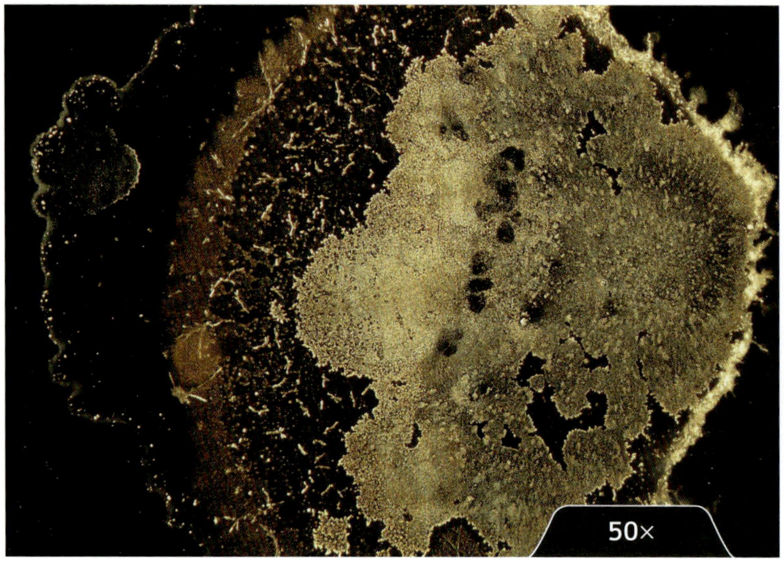

NONORGANIC

ORGANIC

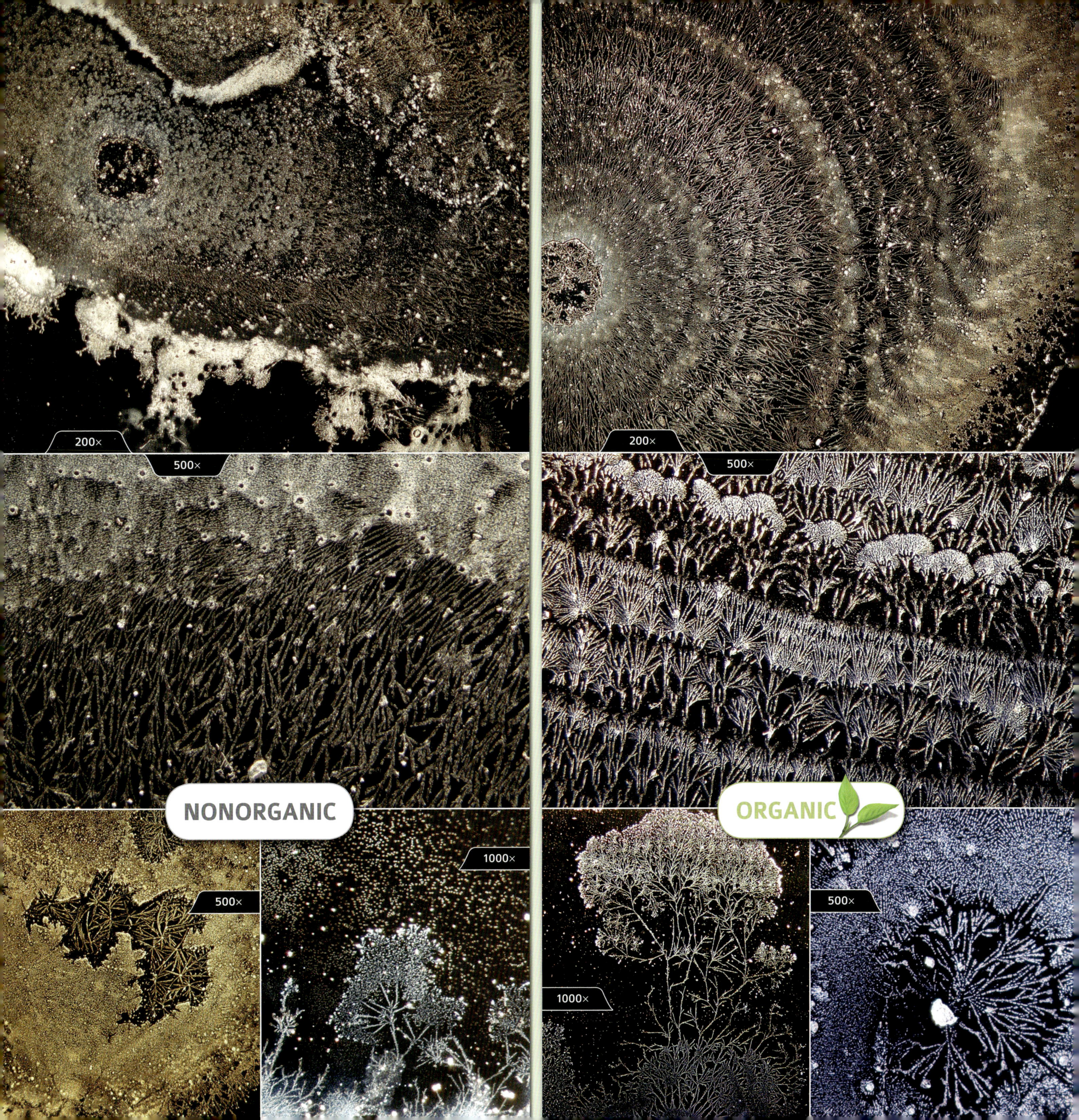
200×
200×
500×
500×
NONORGANIC
ORGANIC
500×
1000×
1000×
500×

250×

NONORGANIC

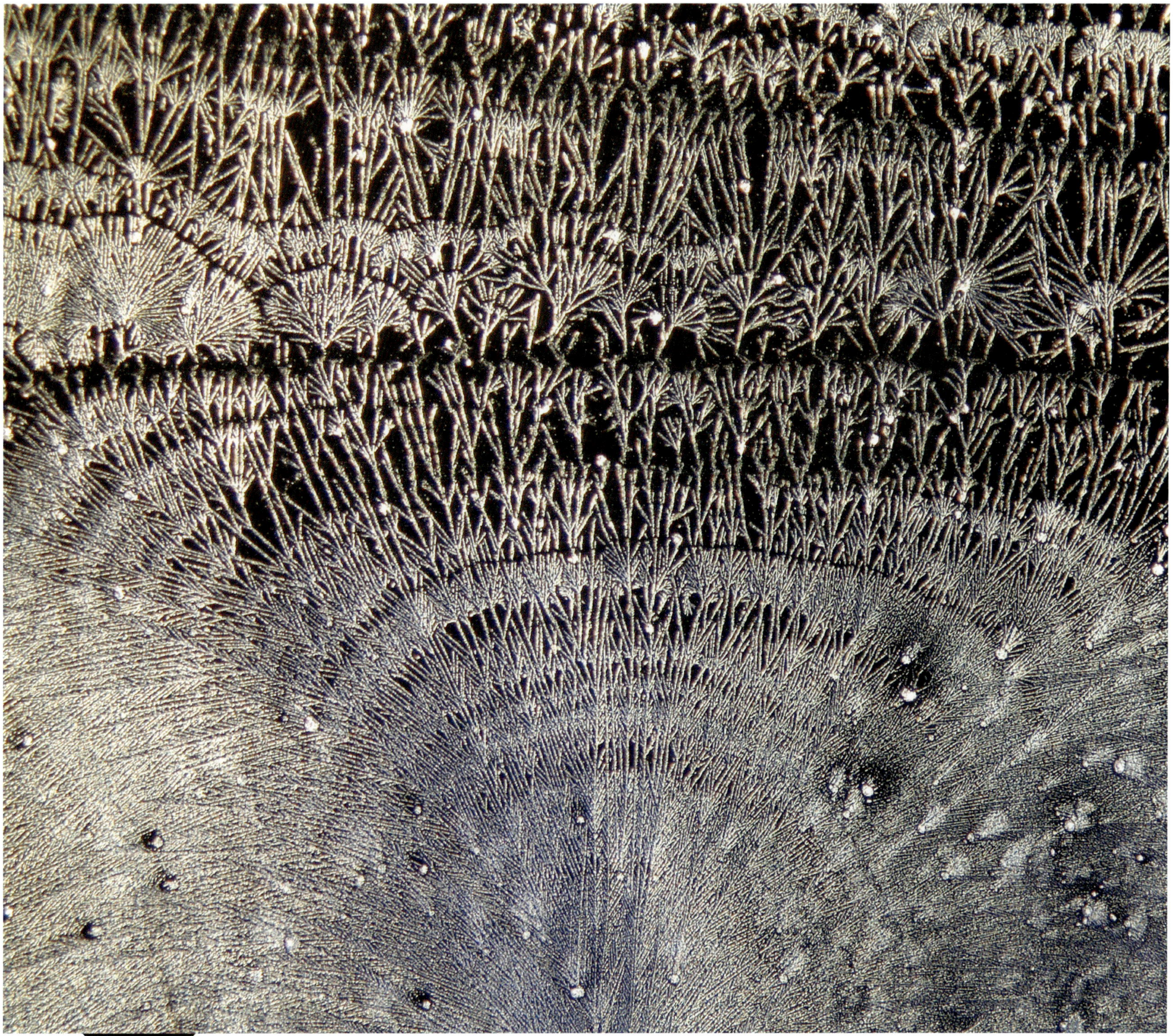

500×

ORGANIC

Grapes

The organic grapes reveal a nature that sometimes forms peaceful, flowerlike tapestries of artistic subtlety or performs a spirited and exhilarating dance with shining blossoms: a joyful love of life, which can be transferred into a wine to grace it with special qualities.

The nonorganic grapes are not only weakened, but their order force (life design principle) is so severely affected that they reveal an image of destruction in the centre, preventing the still available beautiful energy on the edges from expanding.

The reason may be that nonorganic grapes are very often intensively sprayed with toxic pesticides.

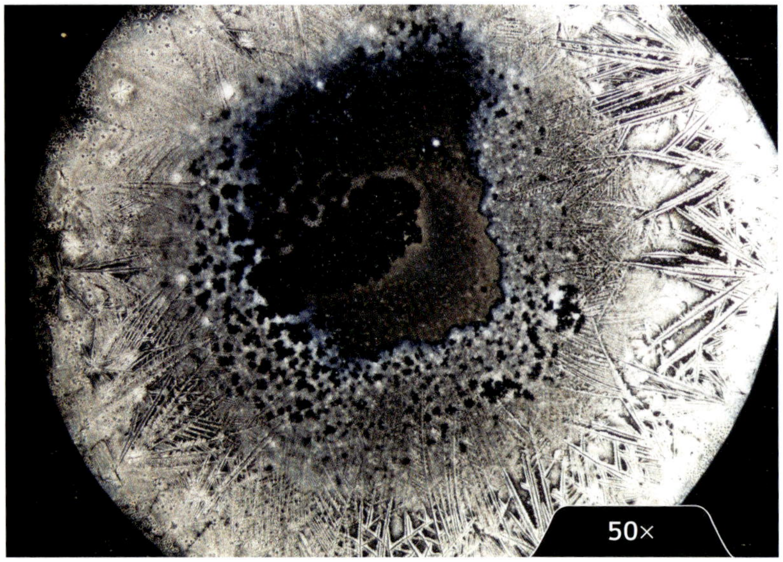

NONORGANIC

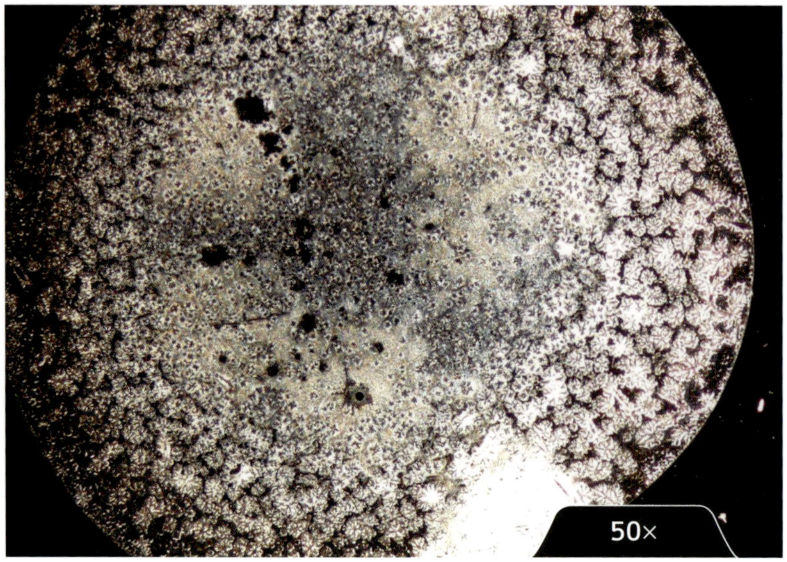

ORGANIC

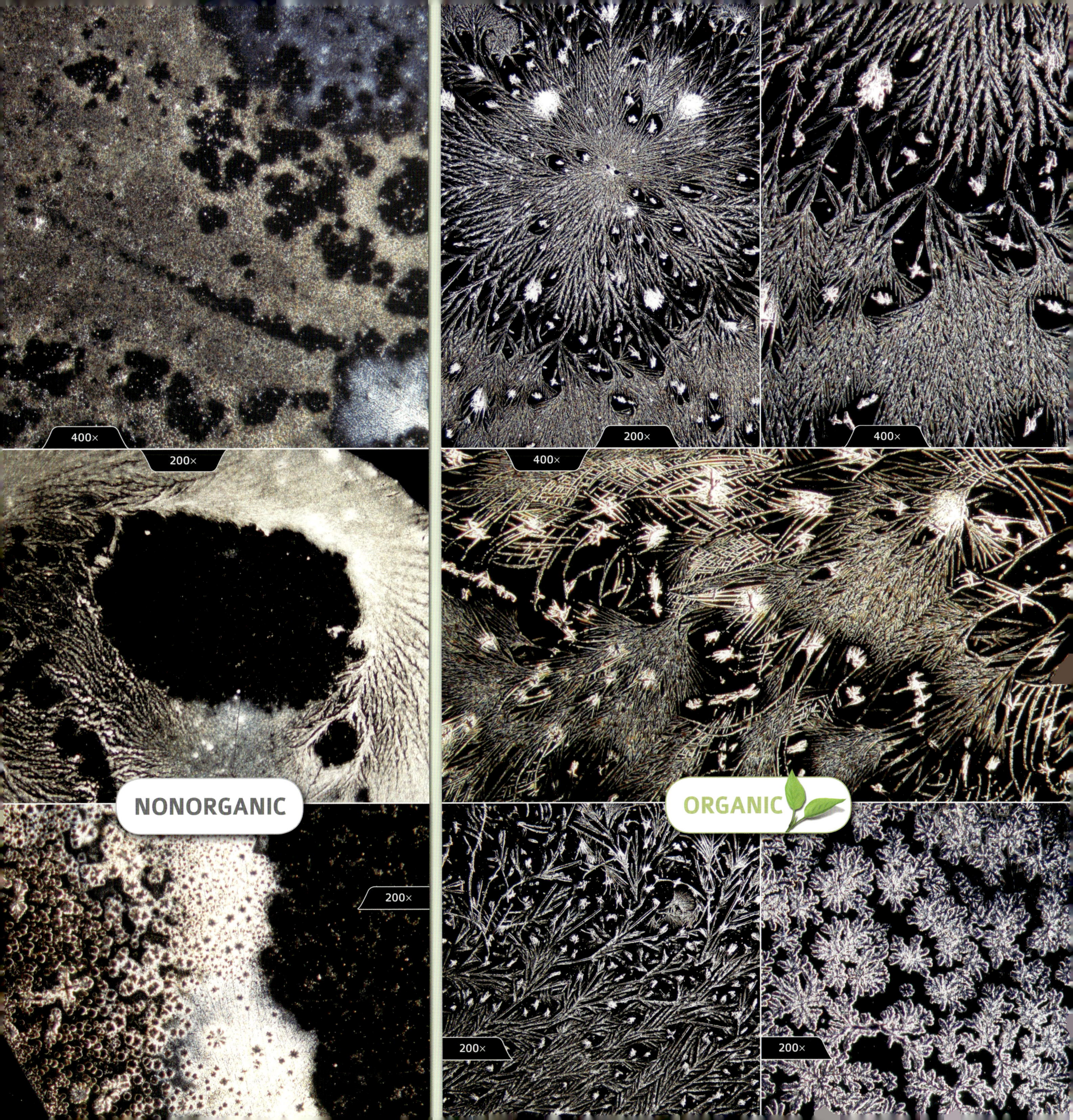
400×
200×
NONORGANIC
200×
200×
400×
400×
ORGANIC
200×
200×

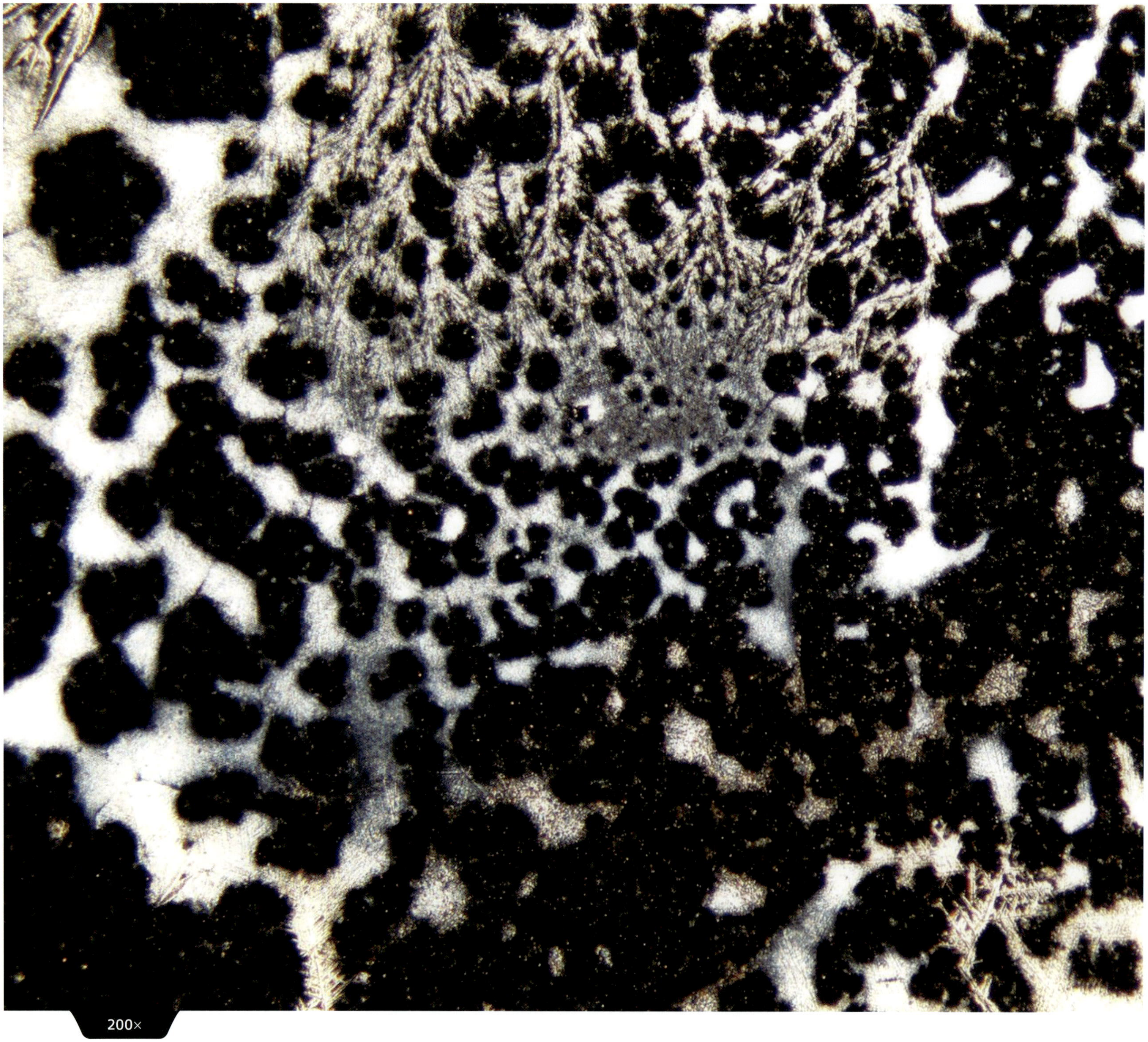

NONORGANIC

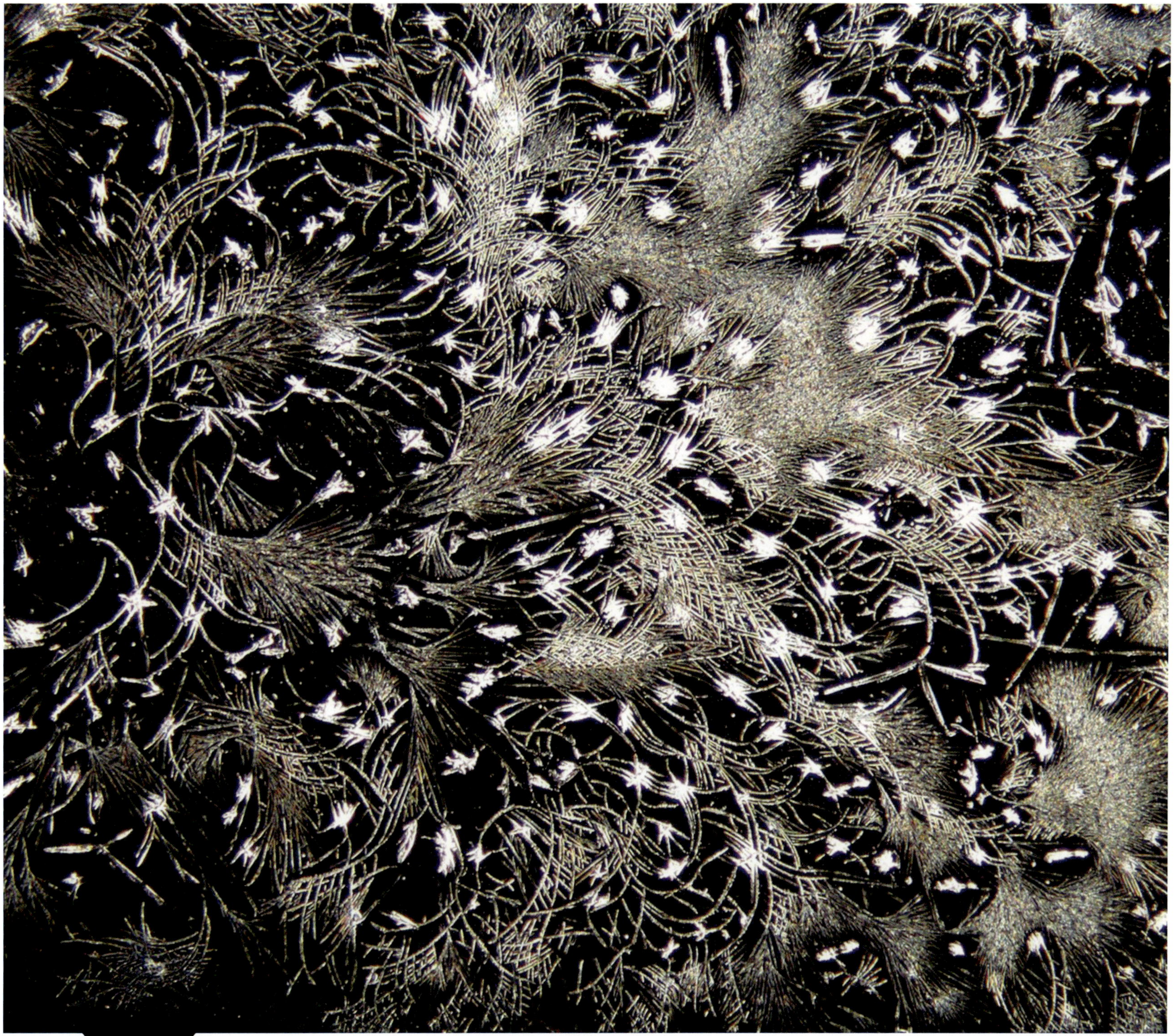

200×

ORGANIC

Orange

Although the nonorganic oranges do not actually show unattractive images, the observer will recognize great differences when compared with the organic oranges: the nonorganic clusters are coarse, broken up, grey and gnarled.

The organic oranges with their rich abundance of filigree twigs and blossoms convey an image of spring, all joyfully sprouting and blooming. Who wouldn't want to decorate the walls of their home with such tapestries and immerse themselves in this sweetly fragrant world of life's natural abundance?

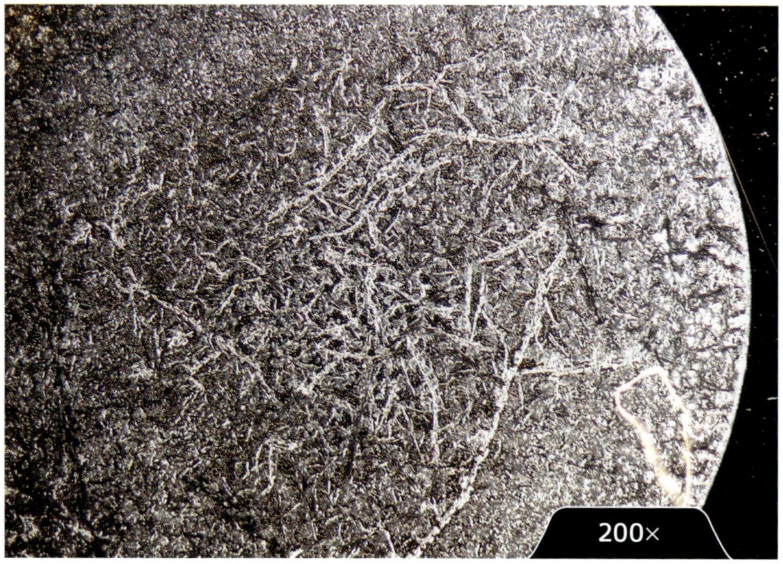

NONORGANIC

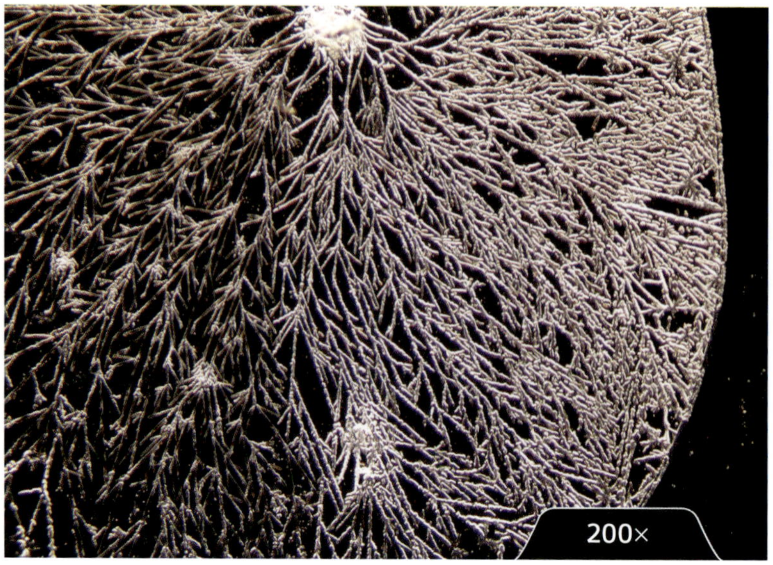

ORGANIC

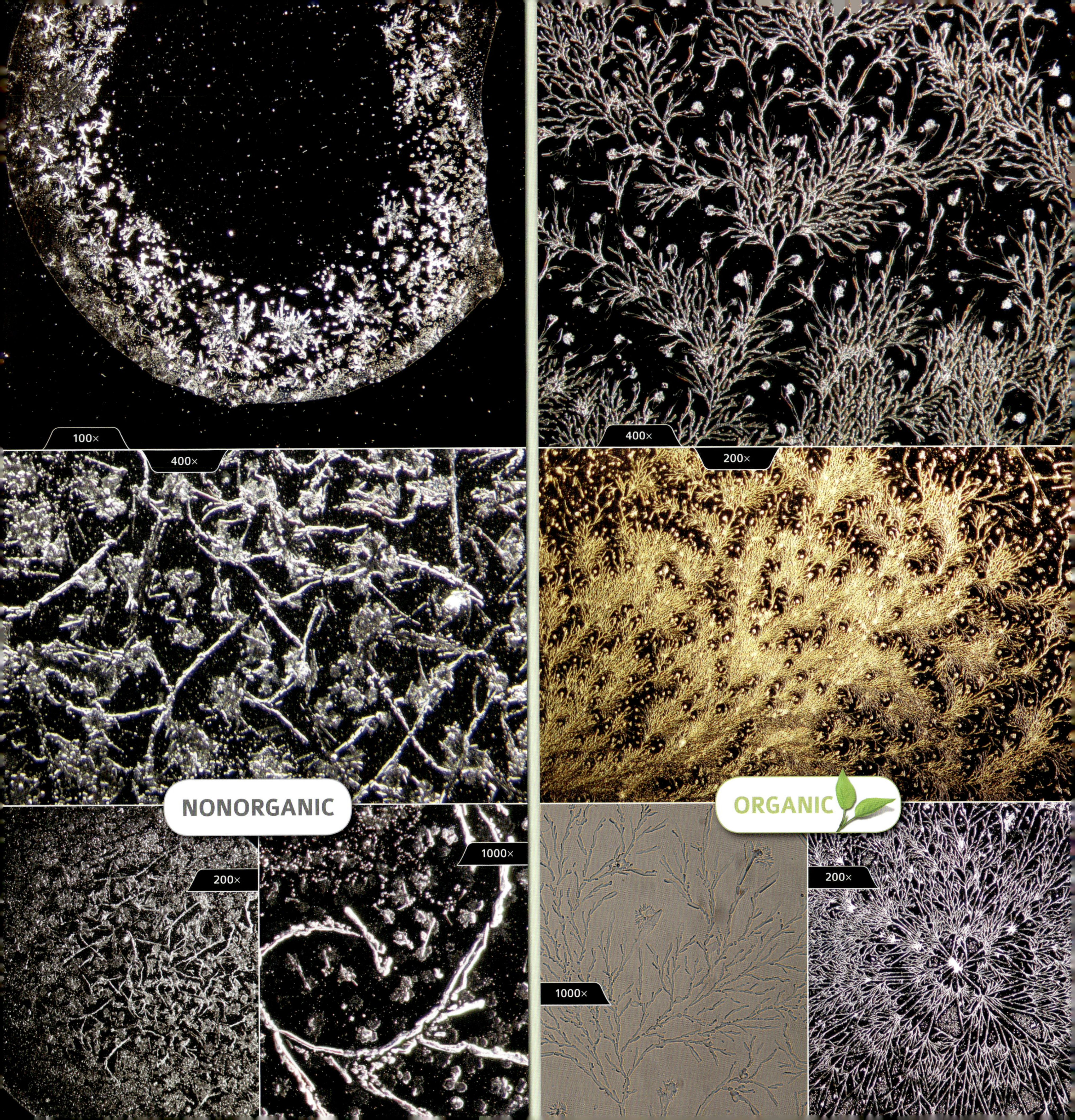
100×
400×
400×
200×
NONORGANIC
200×
1000×
ORGANIC
1000×
200×

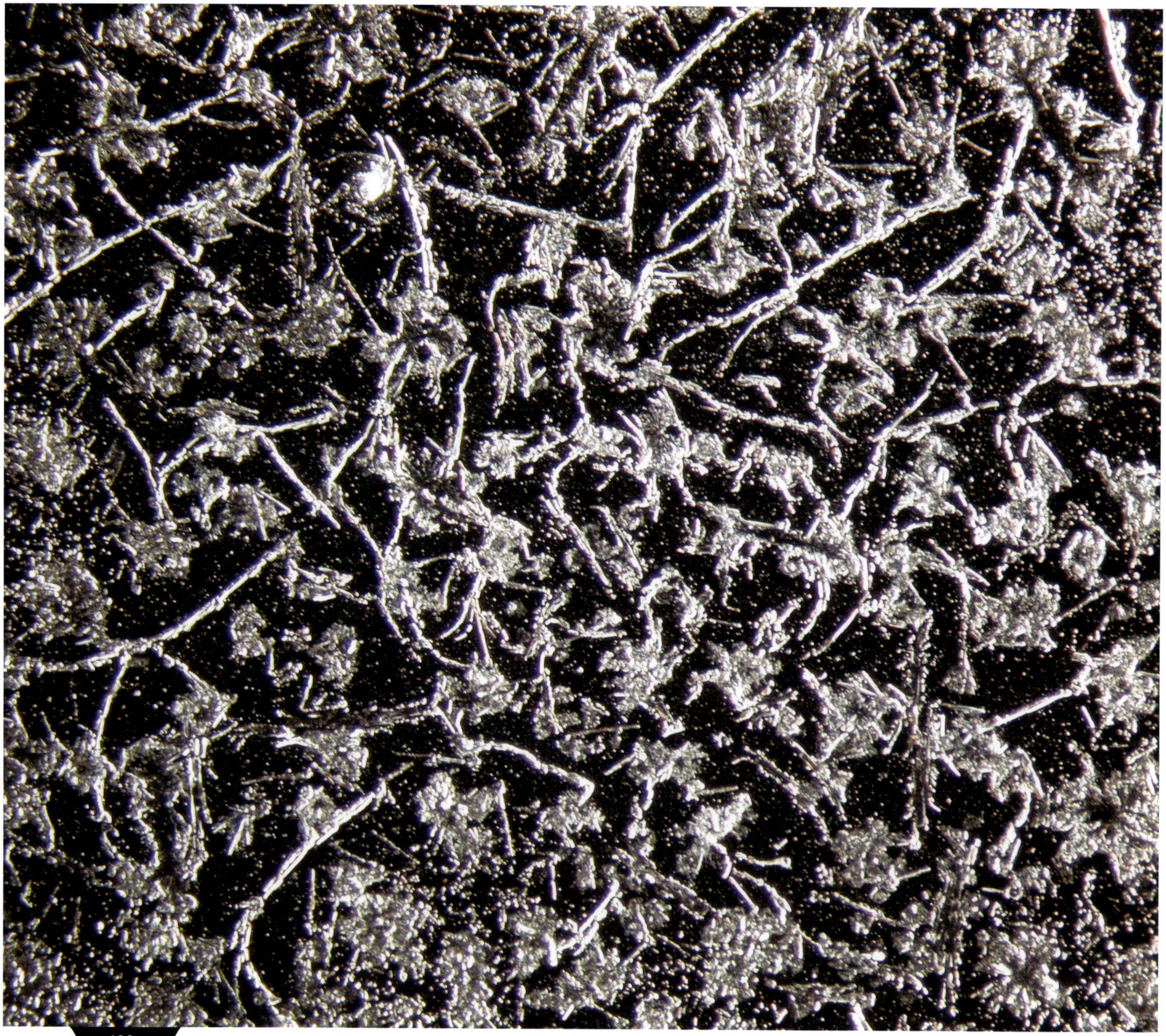

400×

NONORGANIC

200×

ORGANIC

Lemon

Once again the nonorganic lemons present themselves at first glance as rich clusters, yet at a closer look their images of coarse, chopped-up parts full of holes are sadly disappointing. They would like to be delicate and refined, but simply fail to convince.

The organic lemons, however, enchant us with their refreshing effect. Everything sprouts, blossoms and flows beautifully. The distinctive arches remind us of the lemon's buoyancy and healing power, traditionally known to naturopathic physicians who use organic lemons and their juice for many applications.

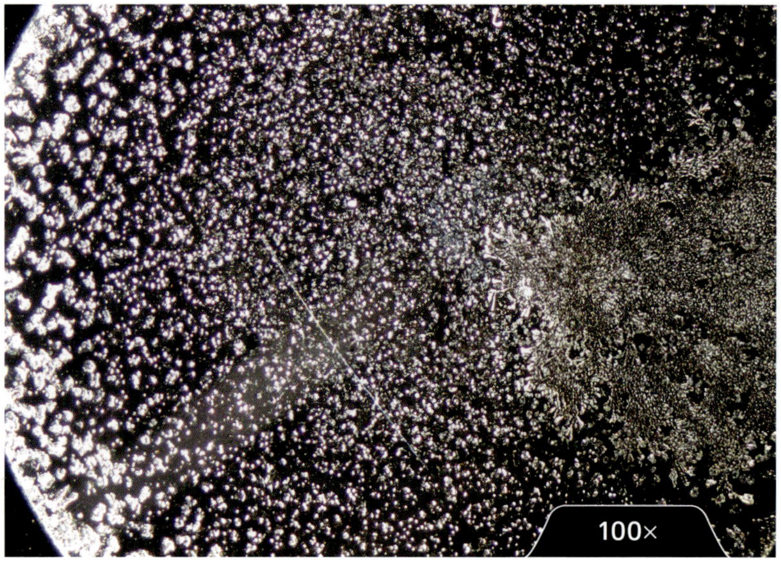

NONORGANIC

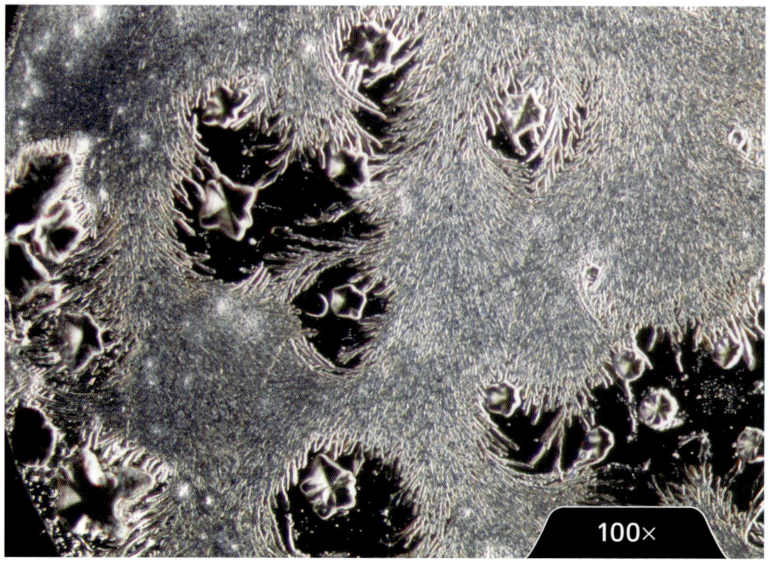

ORGANIC

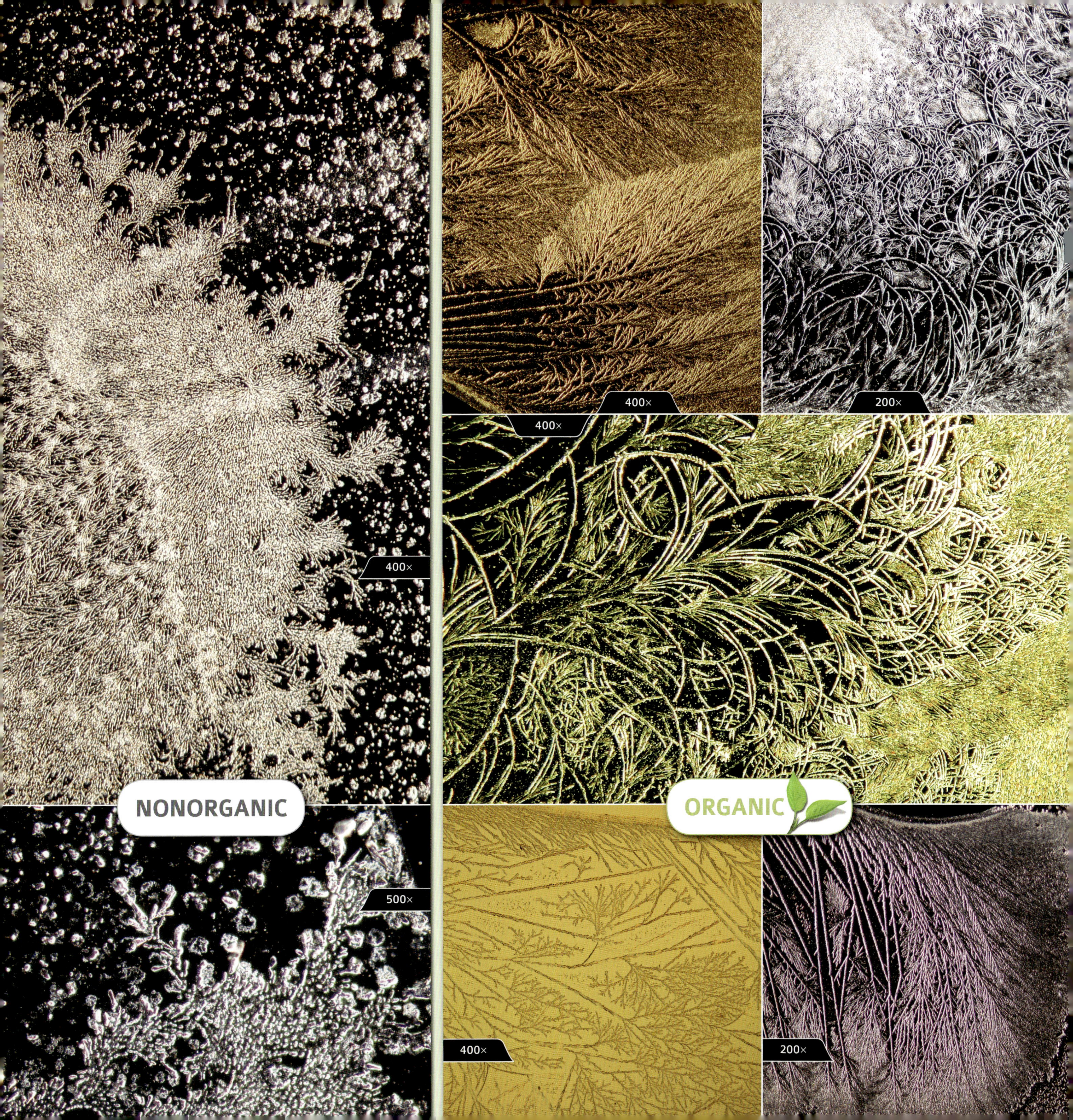
400×
NONORGANIC
500×
400×
200×
400×
ORGANIC
400×
200×

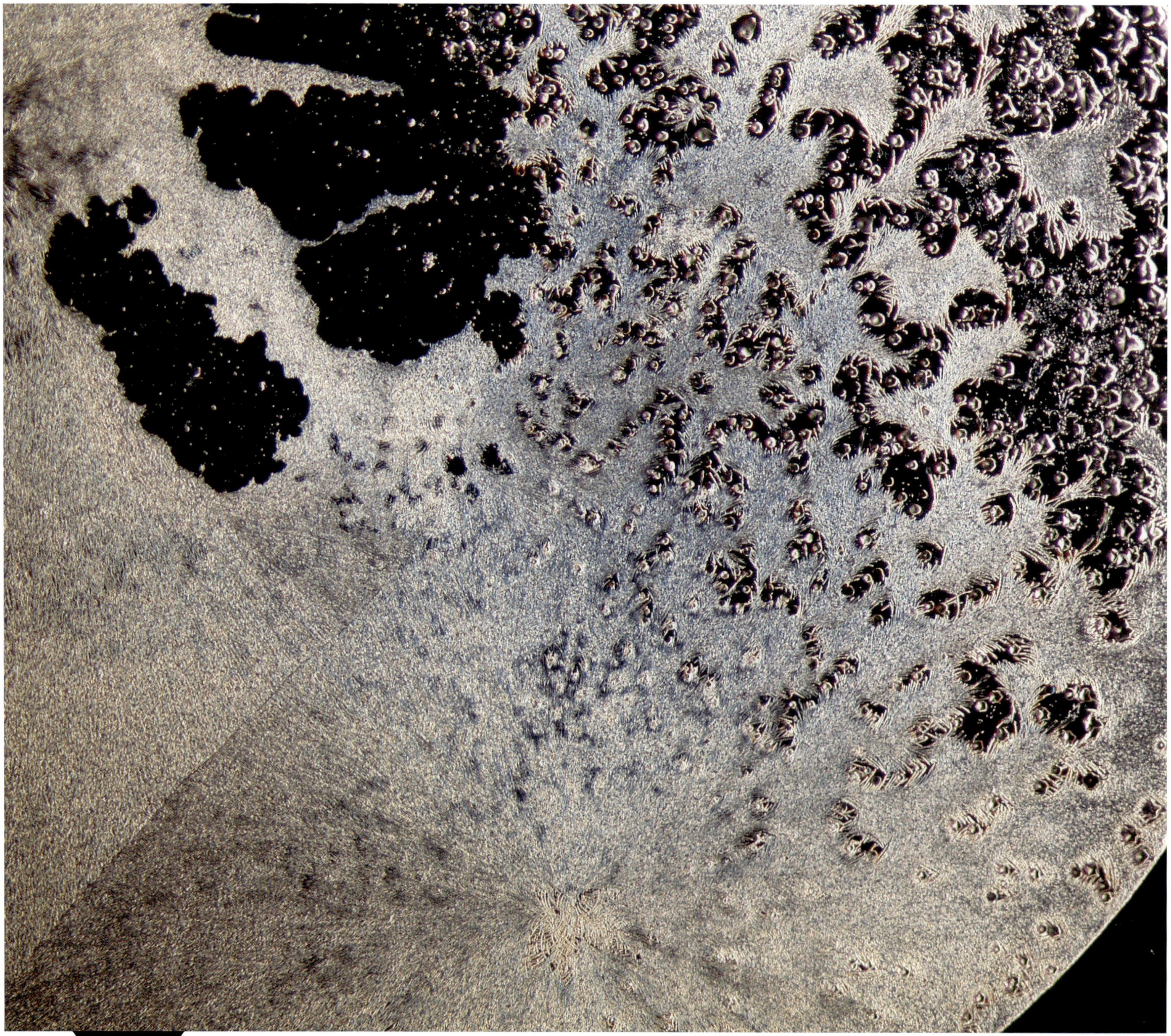

400×

NONORGANIC

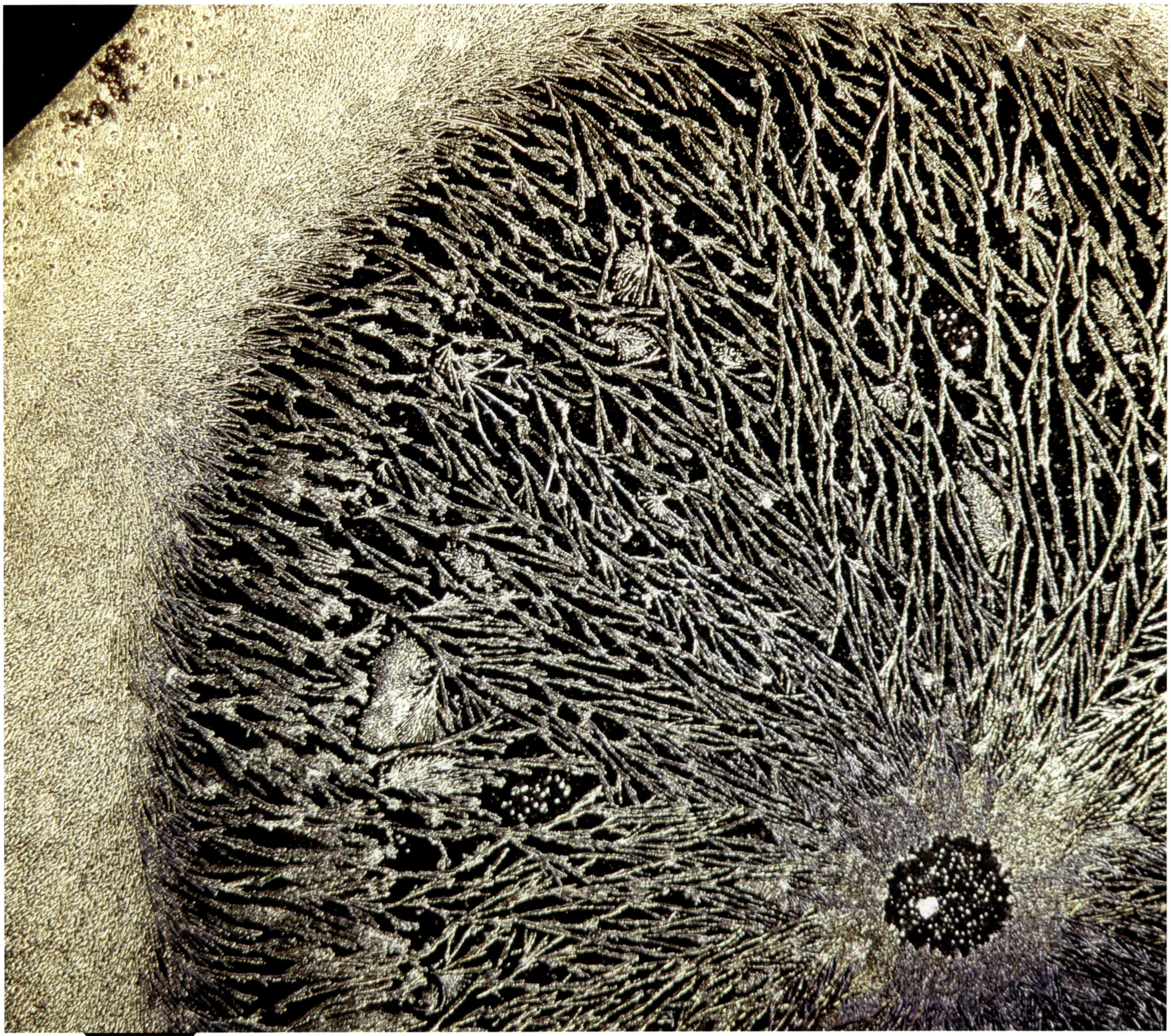

400×

ORGANIC

Persimmon

With its gaping wounds and fragmented fields the nonorganic persimmons reveal that their life-energy has endured heavy losses. Nothing much is left from their natural information.

The organic persimmons, on the other hand, with their dense, plant-like, starry networks, express tremendous vitality. Two particular patterns stand out. In the first, miniature stars grow into streams of growing and expanding clusters that form landscapes of connected strips, similar to the organic apples.

In the other the strips are condensed to smooth hills, reminding us of fruitful landscapes from which centers arise with concentric circles, generating tremendous force.

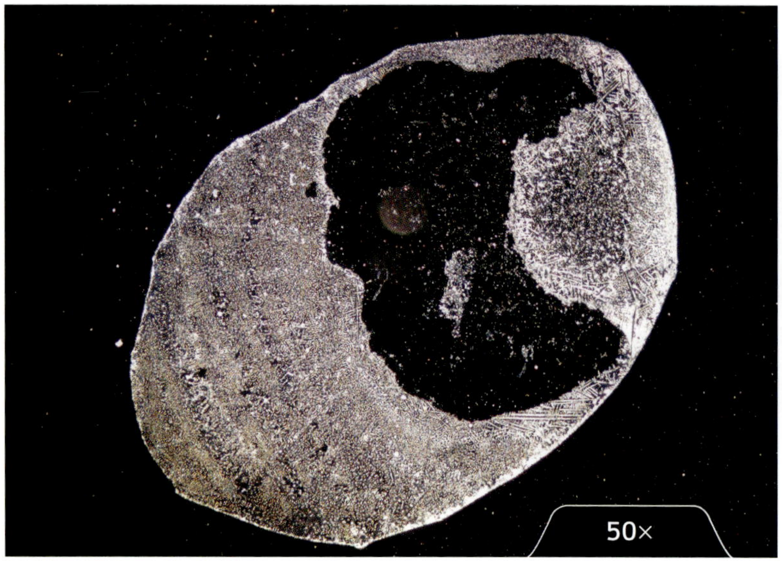

NONORGANIC

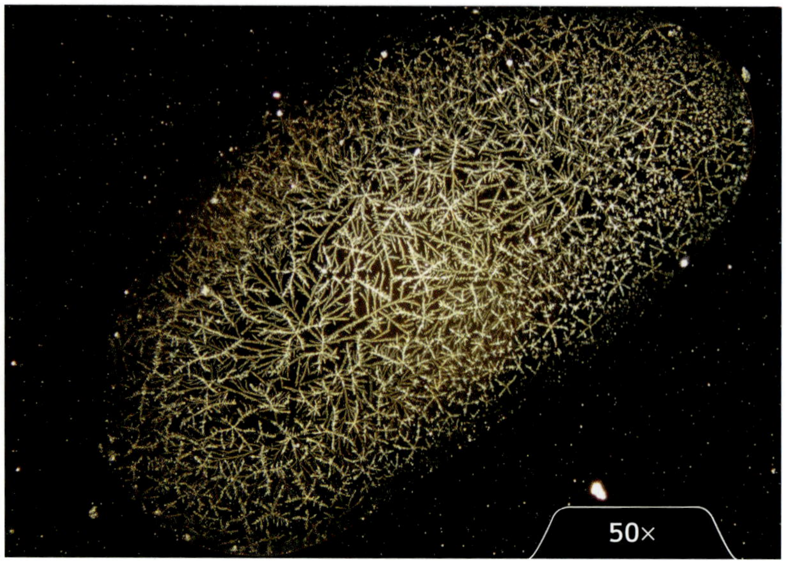

ORGANIC

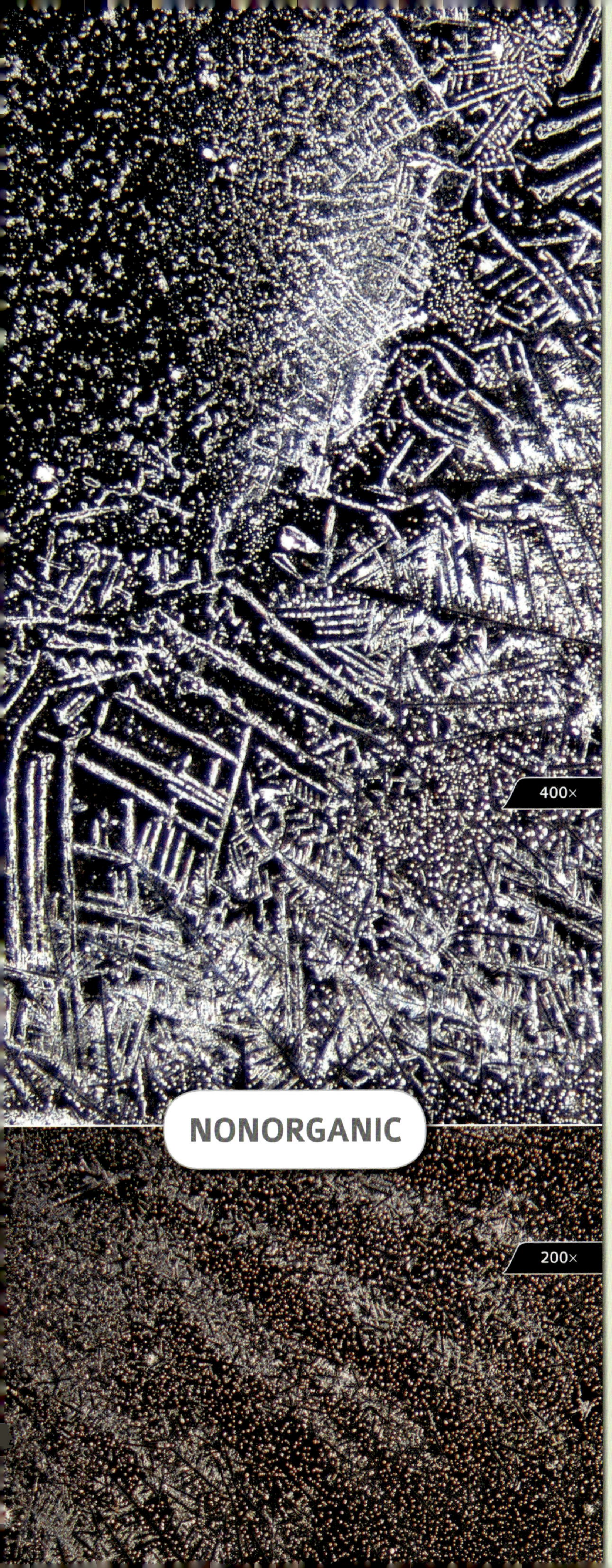
400×
NONORGANIC
200×

200×
200×
400×
ORGANIC
200×
200×

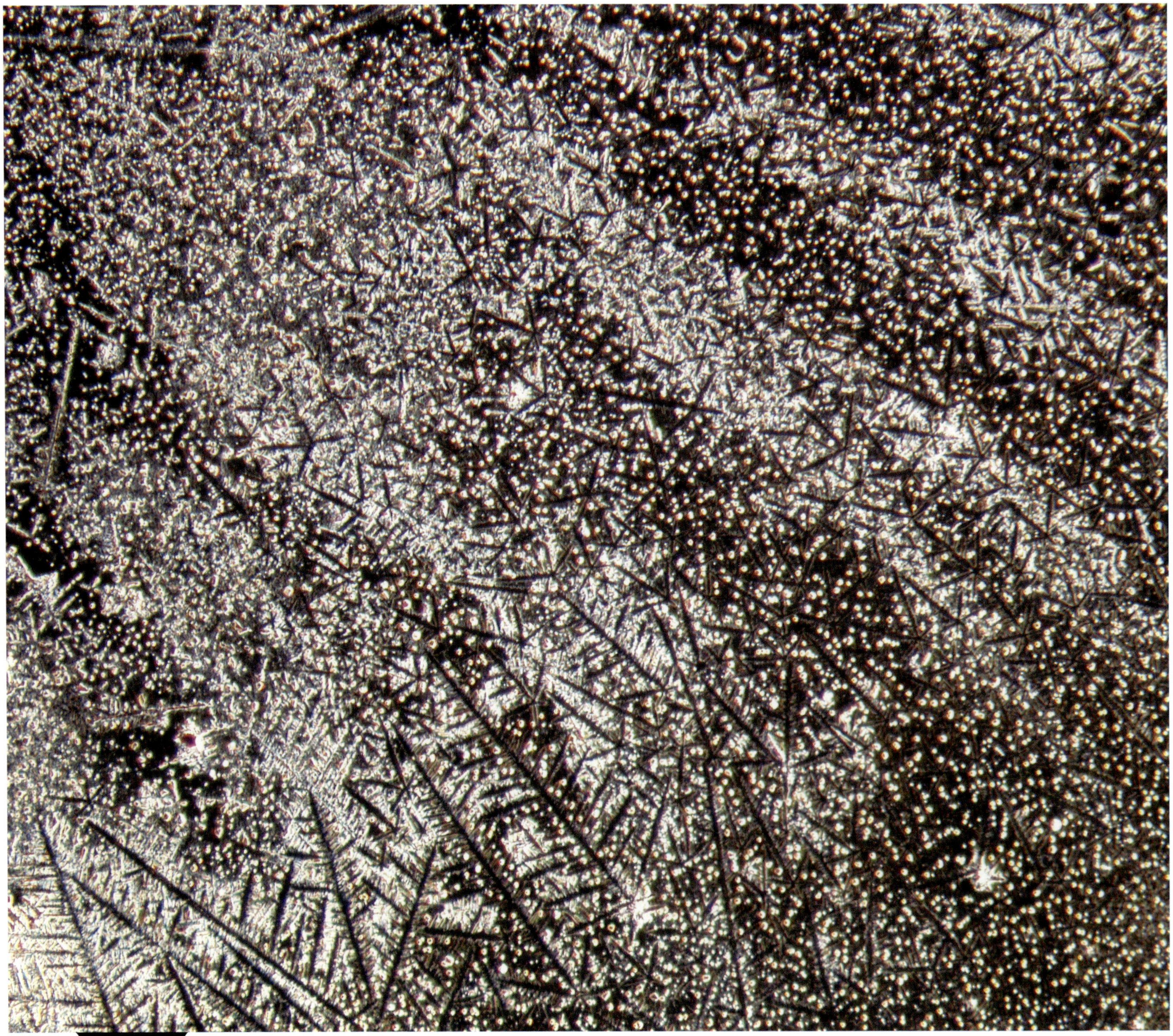

400×

NONORGANIC

400×

ORGANIC

GMO soy drink and organic soy drinks

Soy drink from genetically modified soybeans

The order force of this product is hardly inviting and resembles the rocklike surface of a lifeless, abandoned planet. Everywhere we encounter deserts of terminal moraines and perforated crusts. Only very rarely signs of a former life emerge: sediments of undefined rock or oily drops as industrial waste. In one photograph we encounter a monster, in whose stony mouth a trace of destroyed life drifts by. When we look even closer we can see images that remind us of the desolate retreat of Napoleon's defeated armies in the Russian campaign. In the bigger enlargements we see bent shapes, reminding us of the images stuck in our mind after the tsunami in Fukushima: everything is crooked, branches are defoliated, and everywhere are images of destruction and hopelessness.

Soy drinks from organic soy beans

With a single glance the eye immediately sees the vibrant world of organically growing branches in the BioSoya drink Nature and the full, fresh, lively pulsating atmosphere of the BioSoya drink Indian Chai with its six-petalled blossoms. (Both are called "Swiss Drinks" made by Soyana, available in organic shops in Switzerland and in some organic shops in other European countries.)

400×
400×
NONORGANIC
GENETICALLY MODIFIED
200×
400×
ORGANIC
1000×

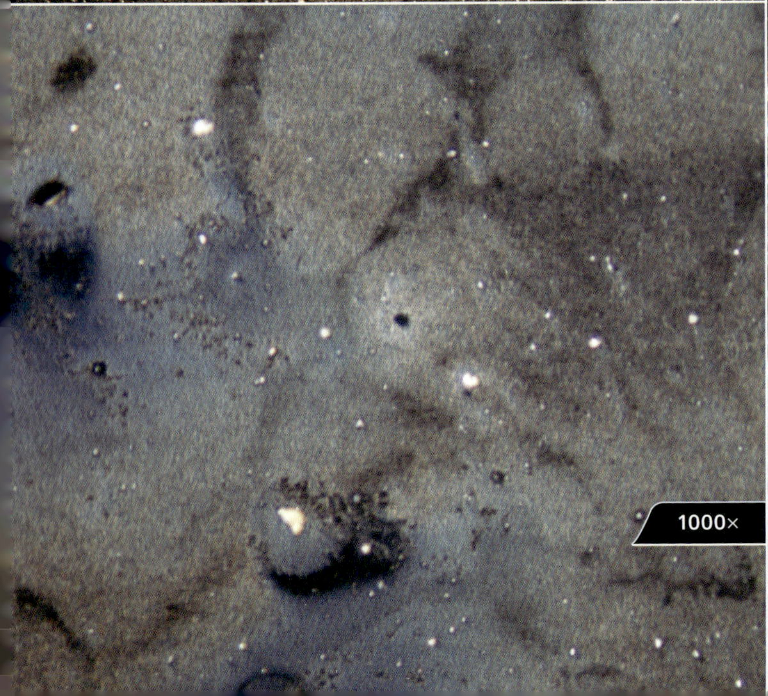

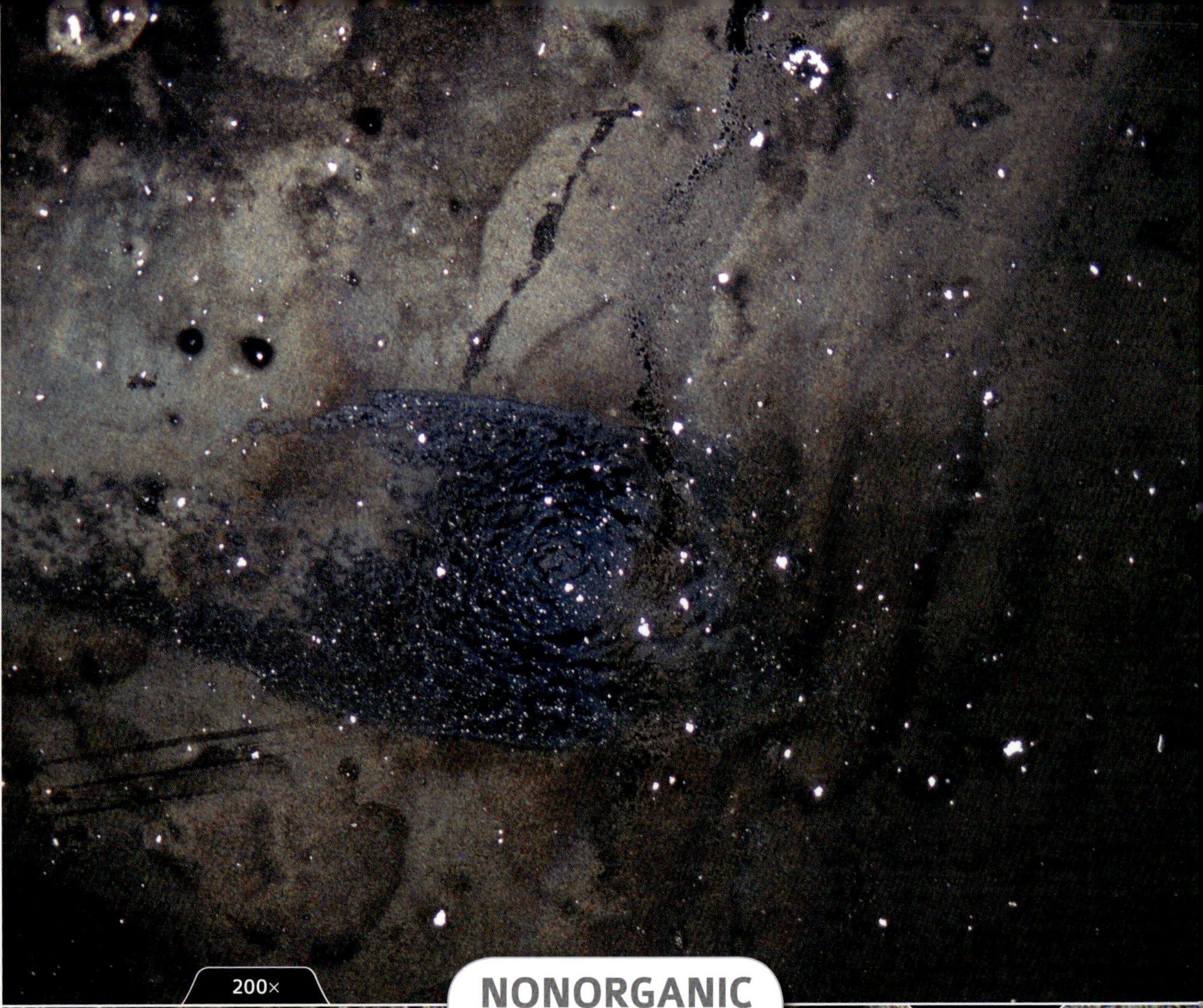

NONORGANIC
GENETICALLY MODIFIED

400×

200×

500×

ORGANIC
400×

200×

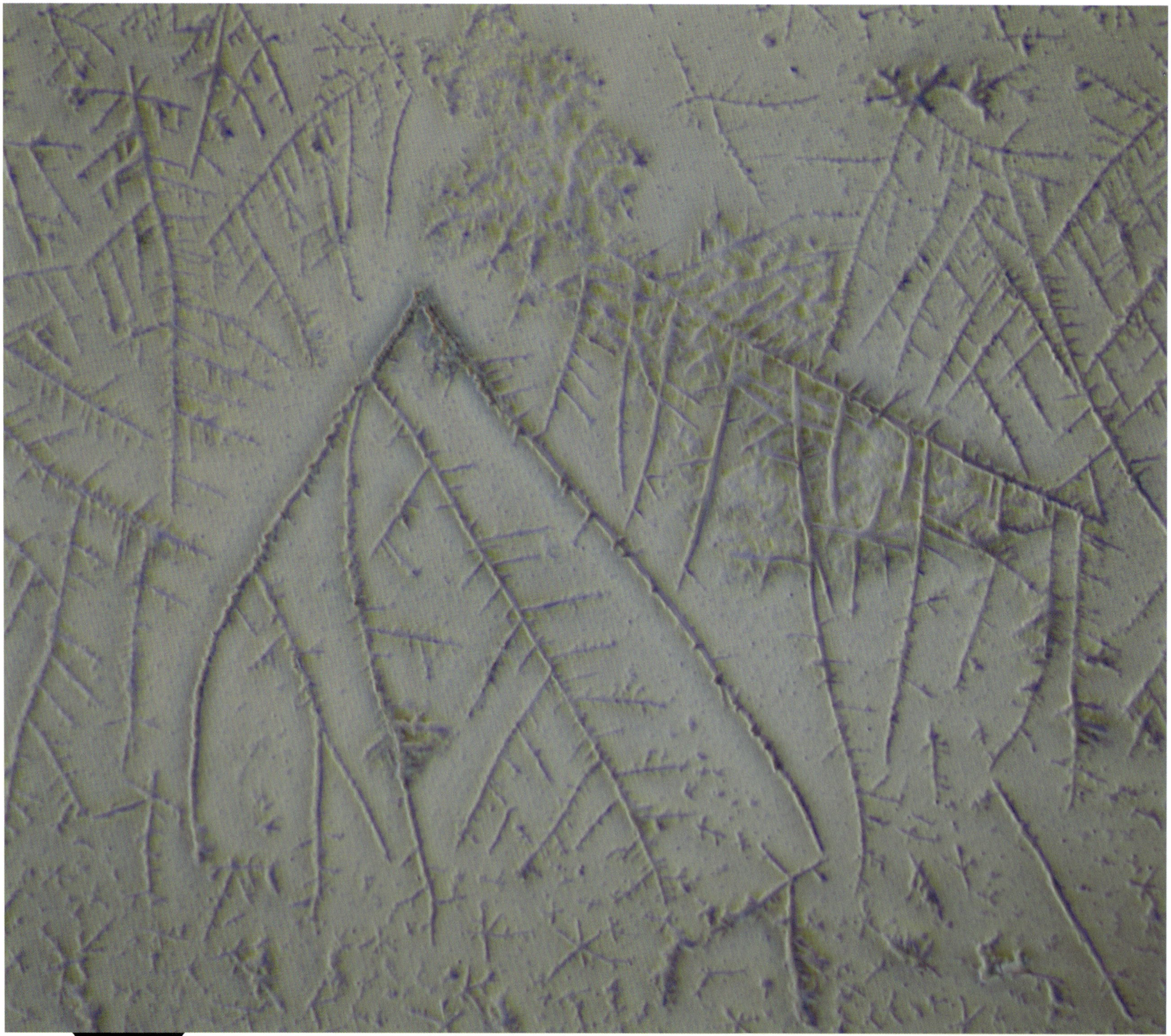

1000×

NONORGANIC
GENETICALLY MODIFIED

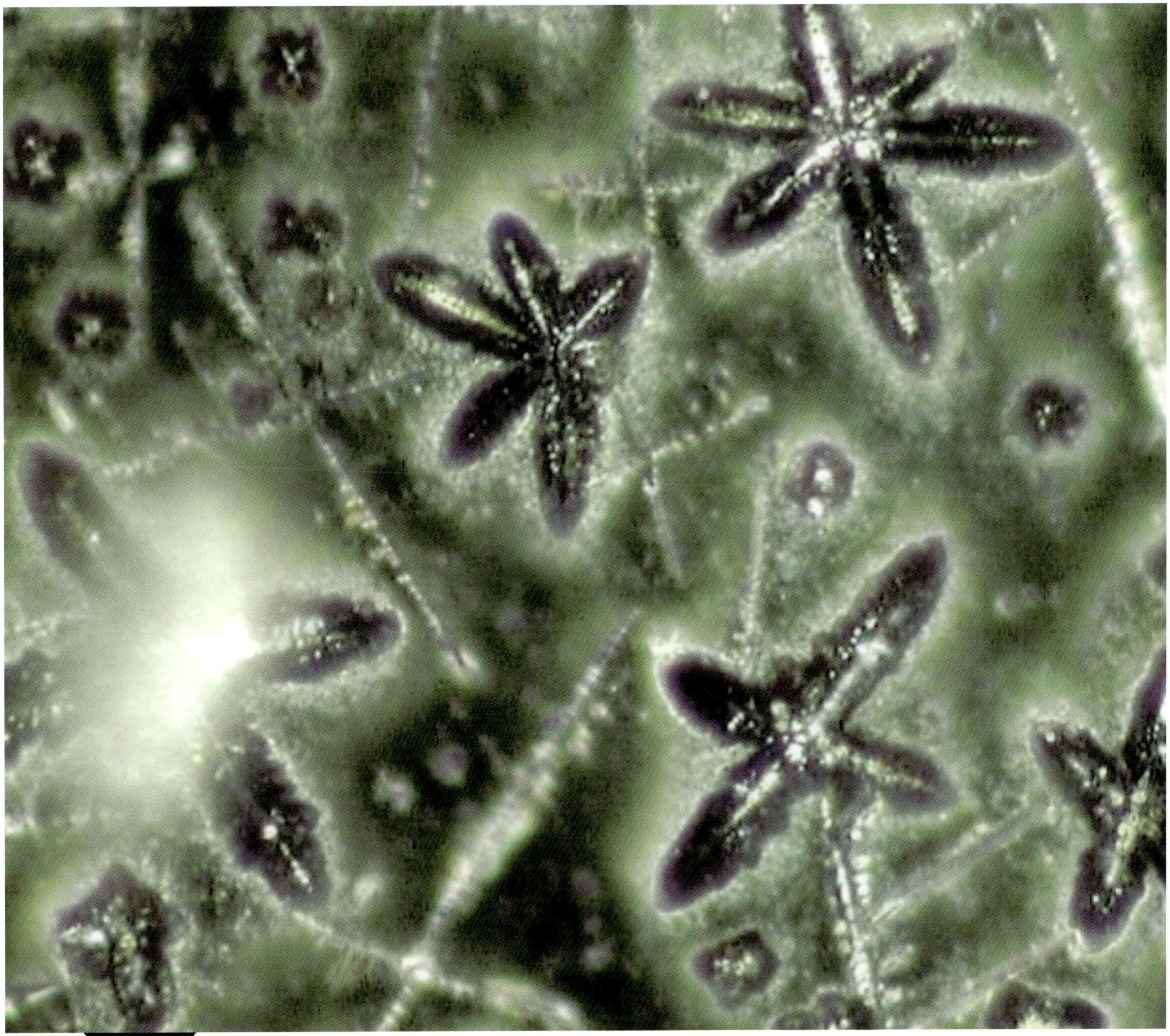

1000×

ORGANIC

1000×

NONORGANIC
GENETICALLY MODIFIED

400×

ORGANIC

Red wine

The conventional Italian red wine researched here shows many impurities. In one picture we can observe beautifully how the order force still is attempting to form a plant. However, the coarse image of the plant, which looks as though it had been scorched by fire, shows that the memory of its natural beauty has been mostly obscured.

The Italian organic red wine by Giorgio Mercandelli reveals finely structured artworks of great creative power and beauty. The highest level of order force (life design principle) fills these images with harmony and peace, with six-petalled blossoms floating among tall plants, artfully plaited natural materials with an amazing ambience and a six-petalled star in a blossoming cornfield.

The last photograph provides insight into the correlation between plant existence and food quality, and makes us aware that organic foods connect us with the landscape in which they have grown. Organic food brings us the consciousness of our natural, organic environment and provides us with an inner connection to living nature.

Eating organic foods thus not only grounds us, but also enriches our relationship with our environment. It is evident that in nature's cradle we feel well.

Could it be that the daily consumption of organic foods grows in us the heartfelt sense of feeling well at home on the Earth? If this is true, then this constitutes the organic foundation of a deep connection to life and life's beauty that in spirituality is called *oneness*. Oneness is the wonderful perception of our all-connectedness. When we are able to feel oneness, we immediately aspire for peace and harmony and also strive to manifest these qualities.

And could it be that we have discovered—in the widespread consumption of nonorganic foods, with their deficient cell-information and lack of order force—a biological cause for the growing indifference and lack of fruitful relationships among modern-day human beings?

When you learn that Giorgio Mercandelli cares for his grapevines as though they were bonsai trees, and even plays music for them, then you will understand that his love is an additional creative force in his winemaking. This is an example of how a human being can contribute to the inner harmony, beauty and power in food, in a way that far transcends the cultivation process as such.

Contact: rilucere@gmail.com
www.riluce.net

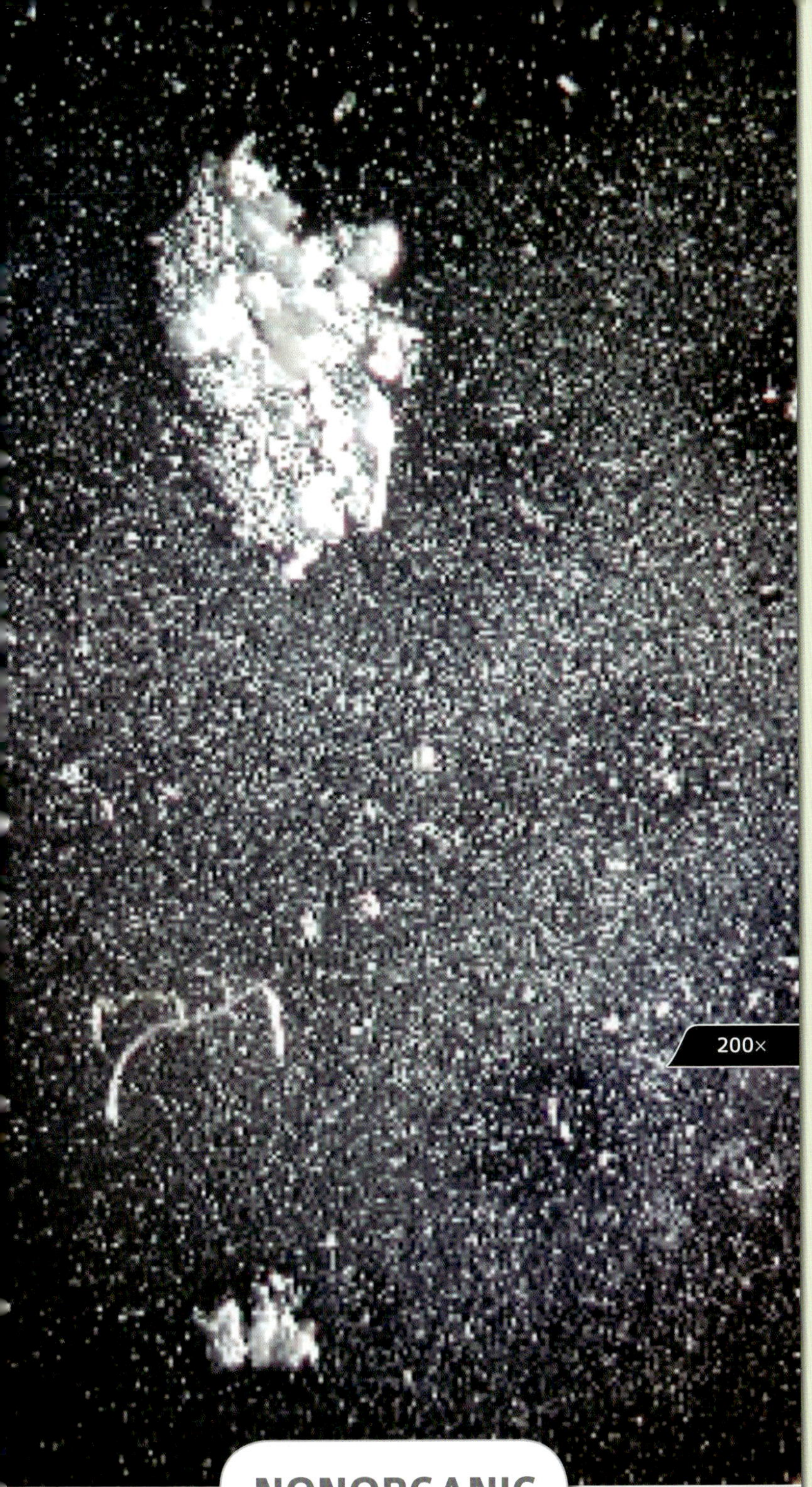

NONORGANIC

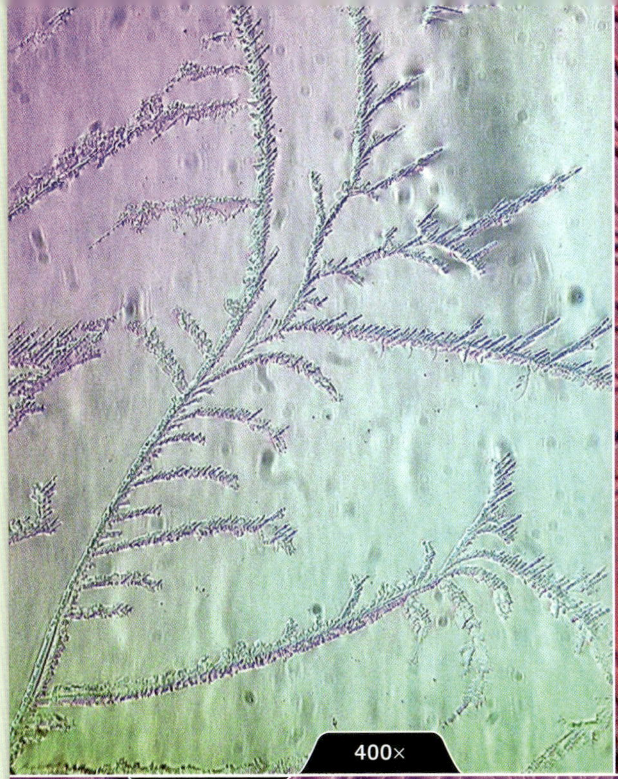

ORGANIC

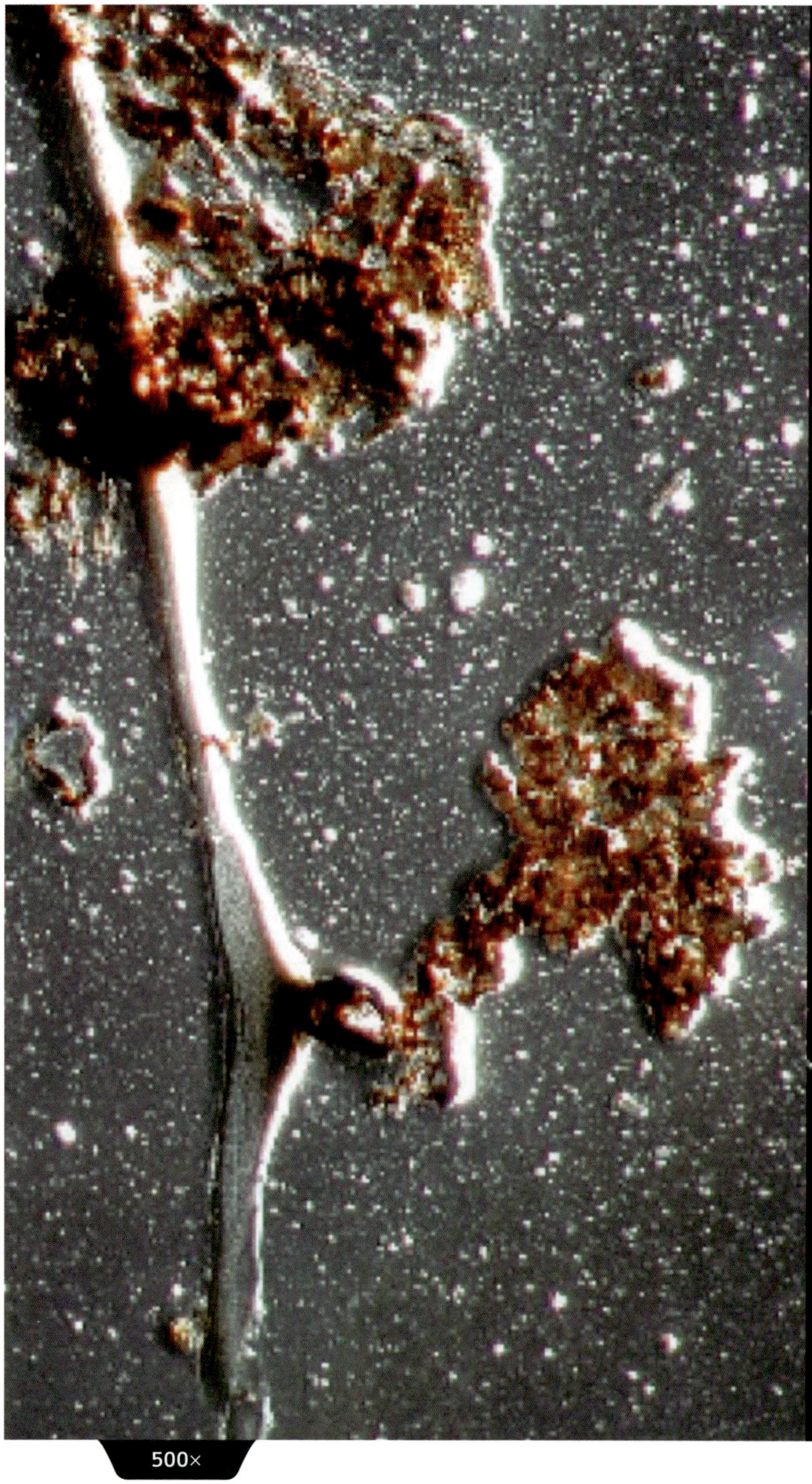

NONORGANIC

200×

ORGANIC

The most striking image shows the native landscape of the grapes: the vineyard *[right page]*. We are just as amazed by this as we were by the images from the organic apples, and a new discovery dawns:

Organic means when the wine remembers the vineyard.

Giorgio Mercandelli practices meditation and also meditates on his grapevines, which is one of the most progressive methods of plant cultivation. Its effectiveness – which may be difficult to believe when it is merely stated as a fact – becomes immediately clear from the LifevisionLab pictures. *(Photo by Lido Vannucchi, Spirito diVino no. 61/2014)*

400×

ORGANIC

Green tea from Japan

The leaves and blossoms of the tea plant represent the most important parts of the harvest. In the image of the nonorganic green tea, although some plant elements are still visible, you will quickly notice that many features are suppressed or have disappeared: the pulse lines are often interrupted without connecting to the clusters. They do not flow in coherence with the plant structure, but instead are fragmented and not clearly defined.

The image of the organic green tea is beautifully developed as an integral whole, and flows with a fulness and strength that immediately transmits a feeling of wholesomeness and balance. We find ourselves completely surrounded by life's abundance. The blossoms are always connected to the branches, the shrubs are powerful, the twigs delicate and precisely defined. The high quality is very convincingly visible and it becomes clear to us why so many people enjoy the healthful influence of this plant.

NONORGANIC

ORGANIC

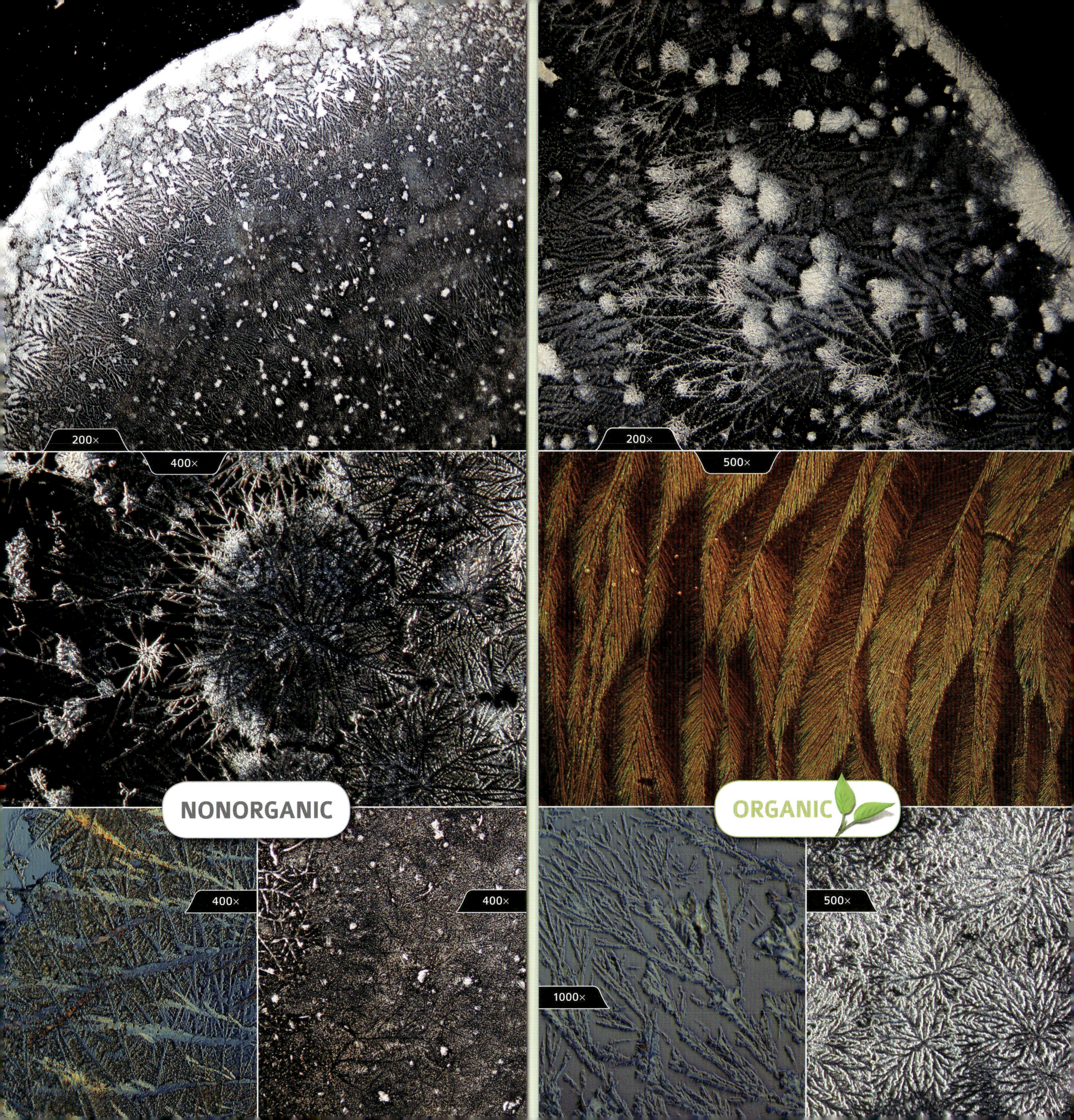
200×
400×
200×
500×
NONORGANIC
ORGANIC
400×
400×
500×
1000×

NONORGANIC

400×

ORGANIC

Coffee

Coffee beans are by nature powerfully built, and so is their life-energy. Although the nonorganic coffee looks similar to the organic coffee, there are some notable differences: the black cut-out forms in the images are much less clearly defined in the nonorganic coffee, whereas they are much more easy to recognize in the organic coffee. In the nonorganic coffee the brightly shining cut-out forms are coarser, less coherently shaped and often they completely disappear or degenerate into a dotted grid or a blurry, rectangular picture like a city plan.

The organic coffee displays more organically grown formations, as well as strong plant clusters that transmit their pulses throughout the entire droplet, forming plant-like tapestries. From time immemorial natural designs have made us feel well and at home. Now, we meet them even deep inside our foods.

The black cut-out forms stand out with their organic shapes and six-petalled flowers. Their shapes connect us with nature: while the big leaves of the black cut-outs soothe us deeply and allow strong emotions to flow, the white drawings have an electrifying quality. The sensory enjoyment we experience when drinking coffee, its rich taste and stimulating effect, is also revealed in the natural formations of its life-energy. Isn't it fascinating to approach the reality of the coffee plant from these different angles and dimensions?

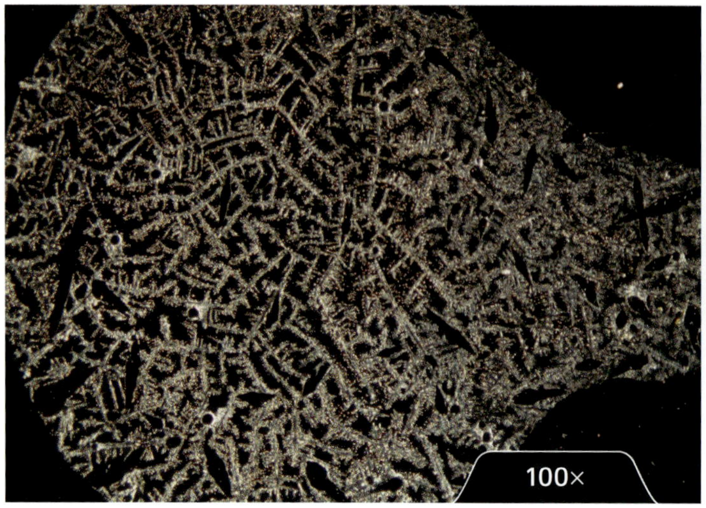

NONORGANIC

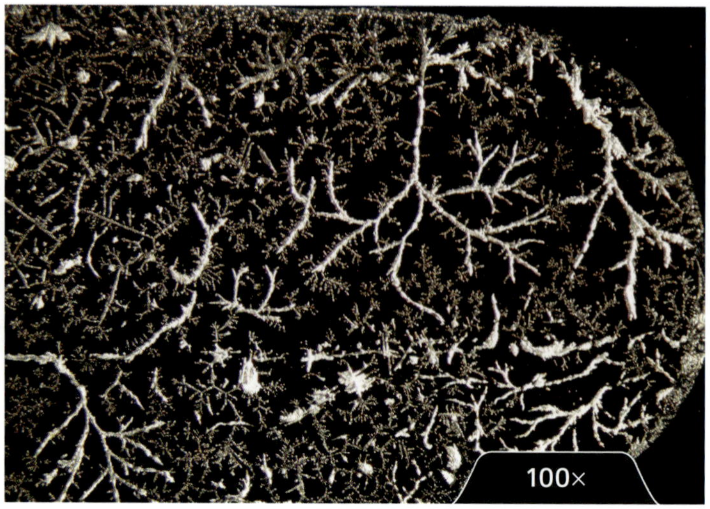

ORGANIC

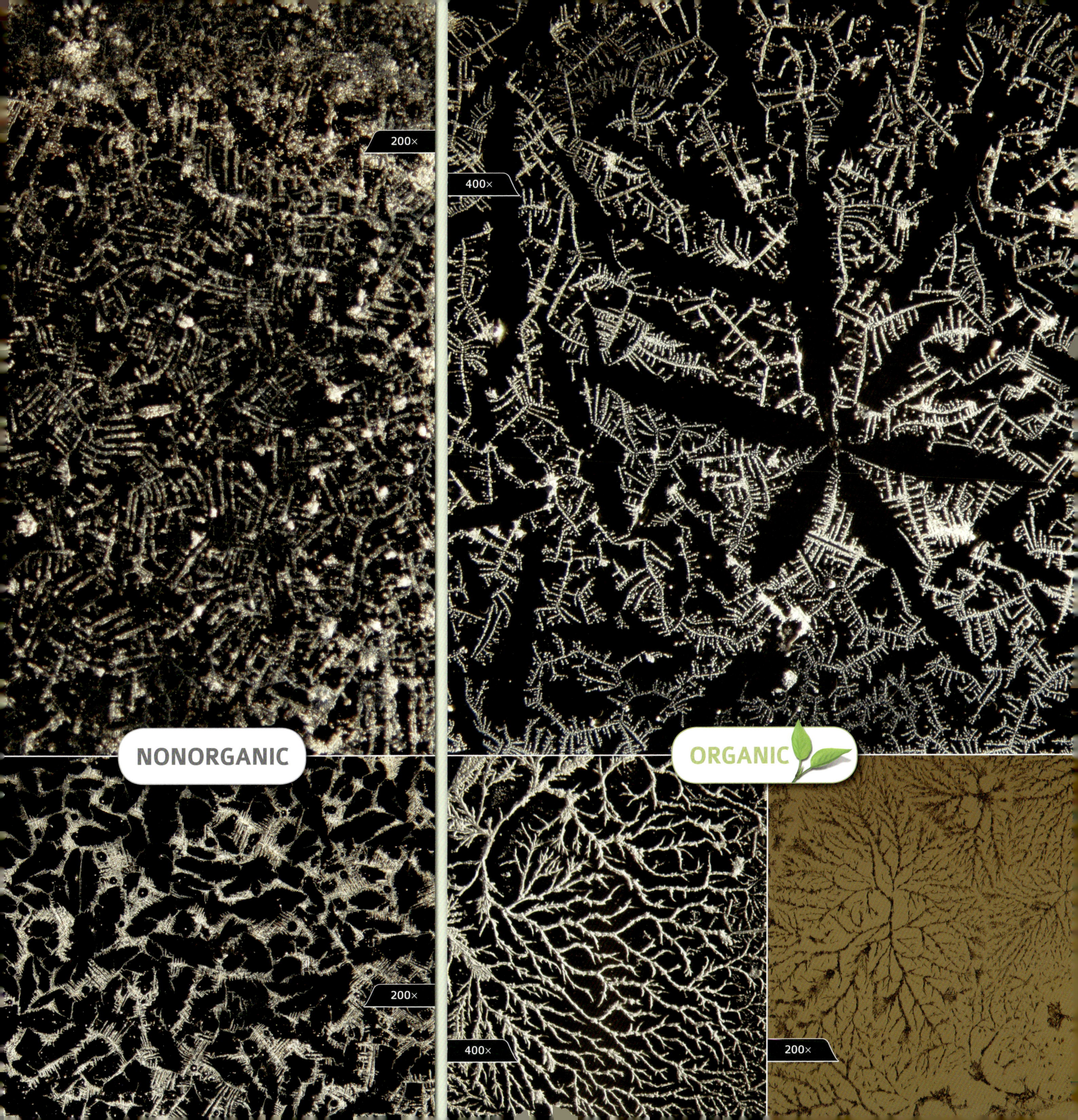
200×
400×
NONORGANIC
ORGANIC
200×
400×
200×

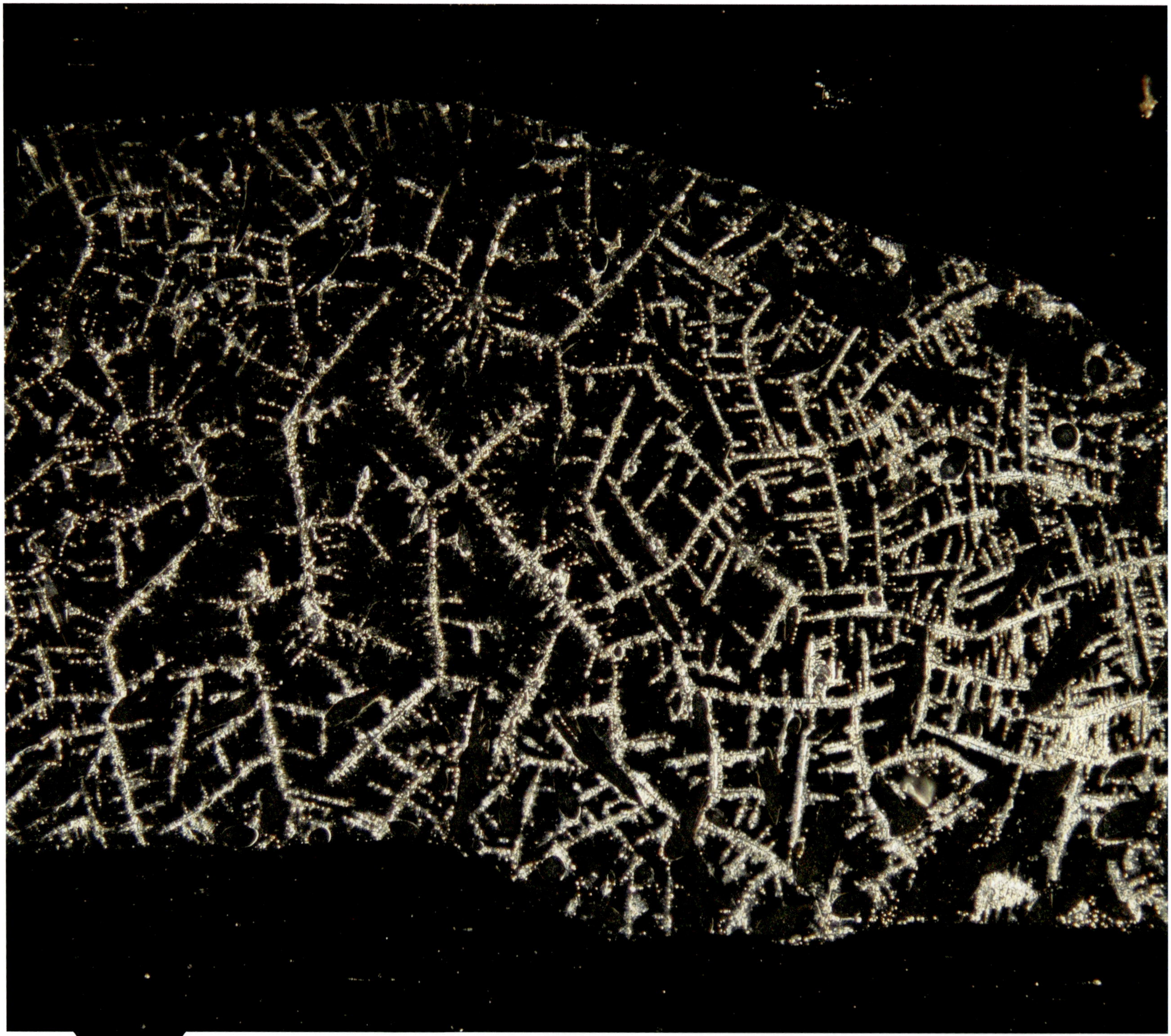

100×

NONORGANIC

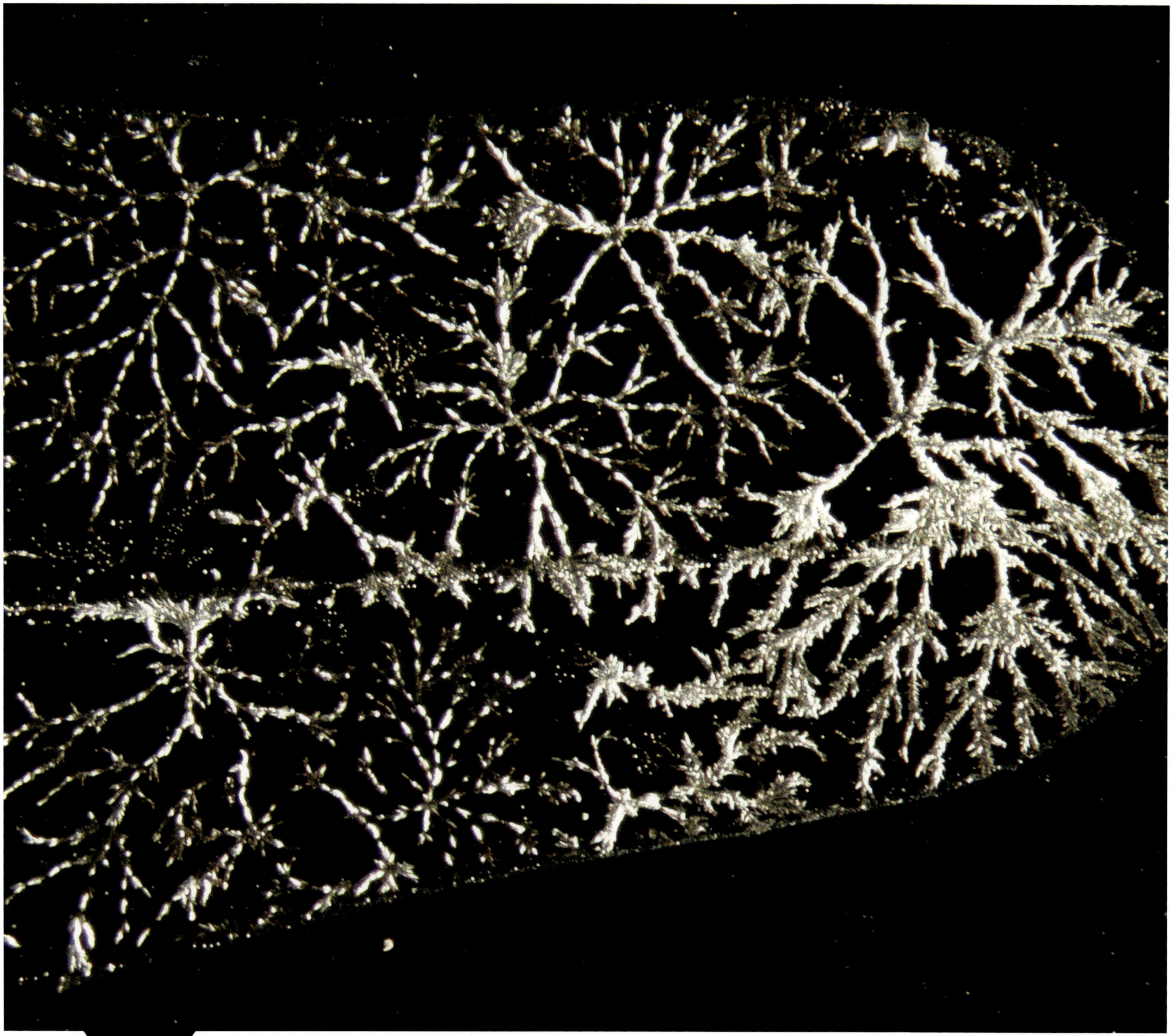

100×

ORGANIC

Cocoa

Cocoa beans are powerful collectors of solar energy. When they can bring to the fore their natural qualities, currents of tremendous power appear.

Much has gone wrong with the nonorganic cocoa: either the droplets are filled with rectangular, industrial-looking forms, or the delicate cables for energy-flow are blurred and disintegrated, sadly failing to transport the energy. The structures are also miserably small, allowing the plant to manifest only a very weak portion of its tremendous potential.

The organic cocoa has a powerfully developed appearance and is filled with solar eruptions that make you feel warm just by looking at them. When observing these images we immediately grasp why we are filled with joy and warmth when we enjoy organic cocoa. Organic cocoa gives us the power of the tropical sun.

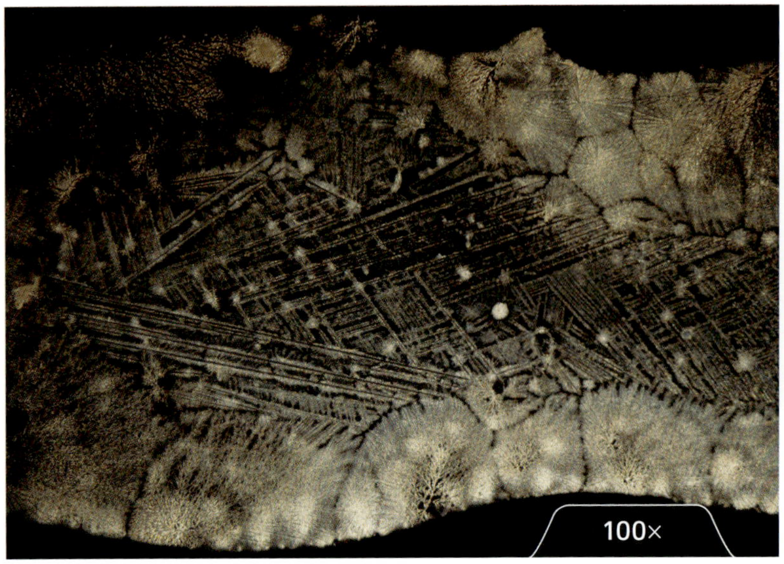

NONORGANIC

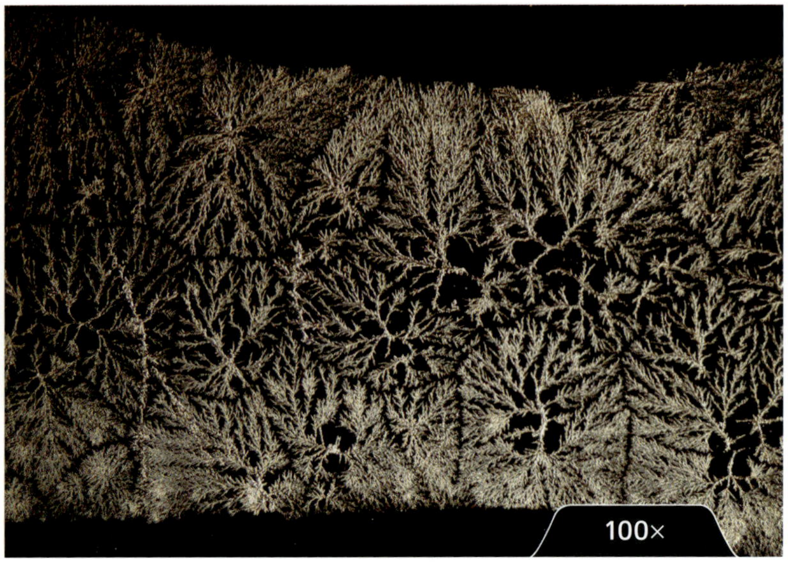

ORGANIC

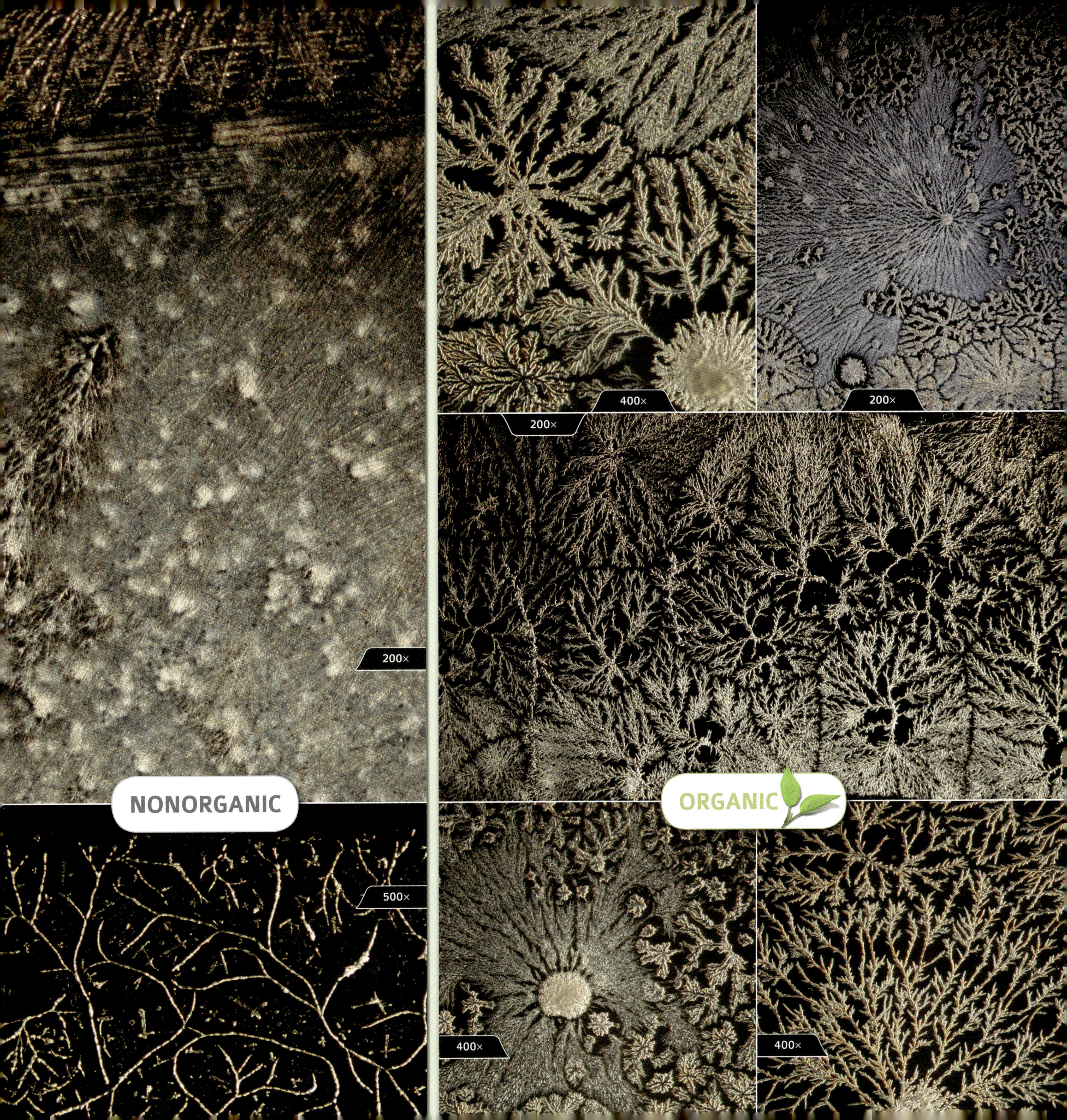
200×
NONORGANIC
500×
400×
200×
200×
ORGANIC
400×
400×

200×

NONORGANIC

400×

ORGANIC

White cane sugar

Almost entirely cleared out, the drops of nonorganic cane sugar hardly show any signs of life at all. Instead, an inner weakness permeates the image: only a few dots remain with some last fragments of memory, before they vanish in complete forgetfulness.

The organic cane sugar fills the drops with characteristically formed plant-like shapes—the close observer will notice the black cut-out forms in between, which appear like moss or vines growing on stone.

These images make us aware that nonorganic white sugar, consumed in vast quantities today, provides nothing but empty calories, sadly failing to offer us any life-energy at all.

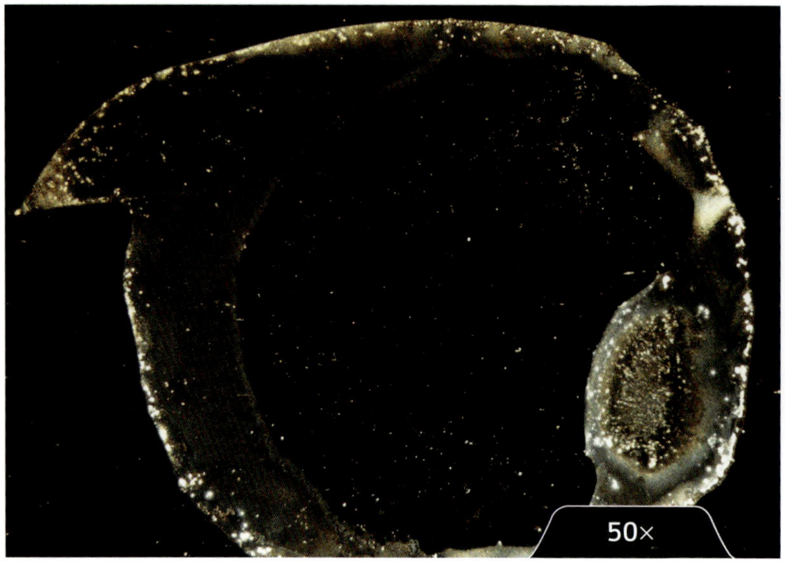

NONORGANIC

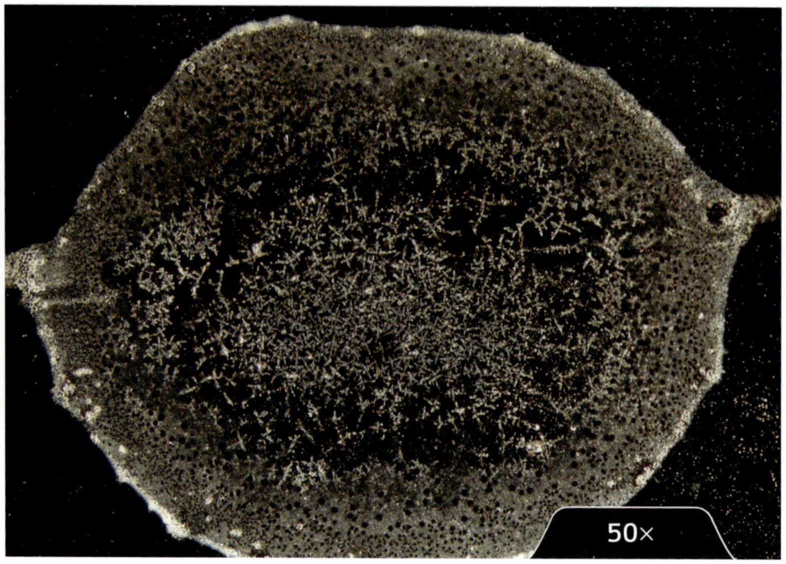

ORGANIC

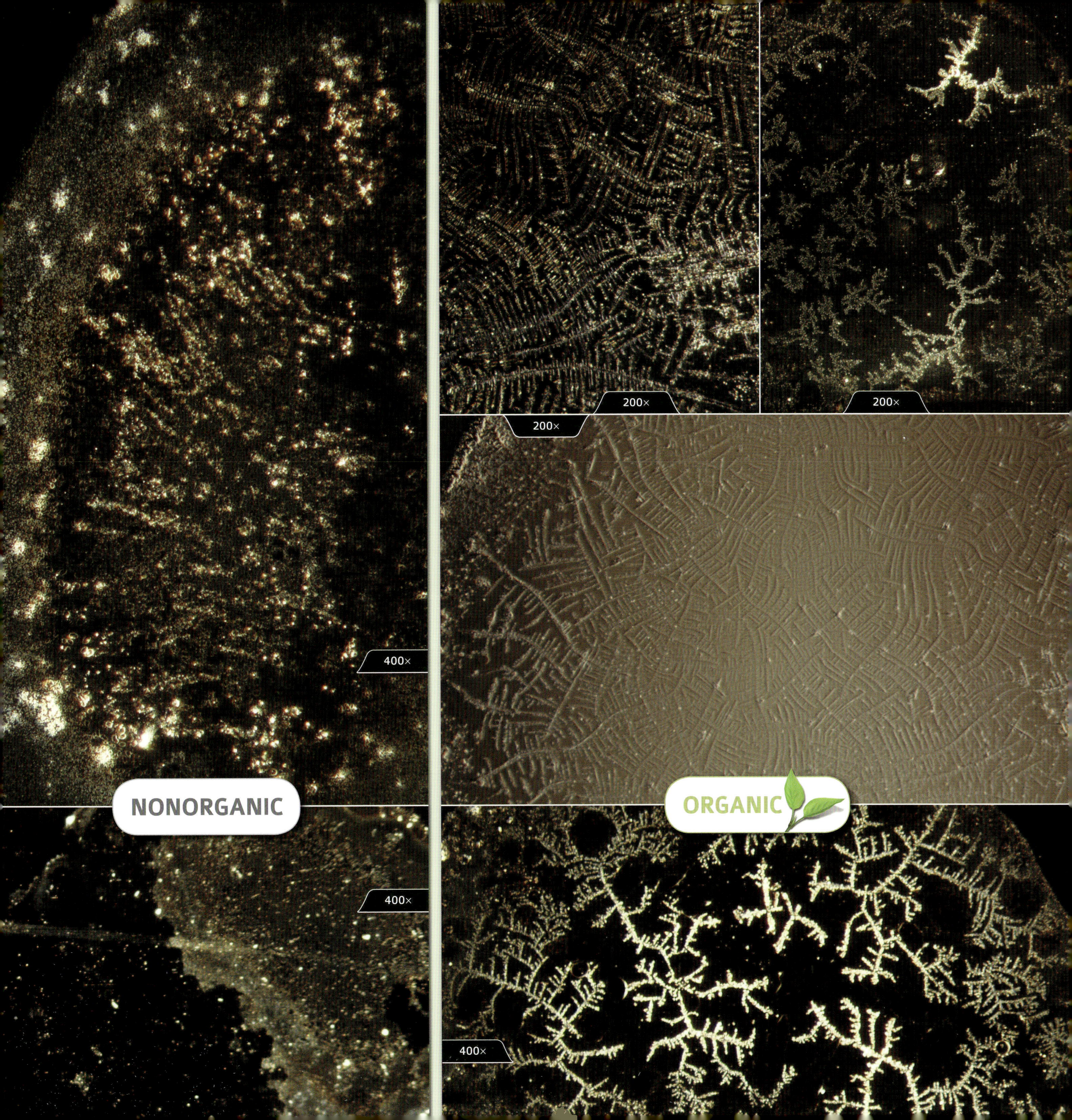
200×
200×
200×
400×
NONORGANIC
ORGANIC
400×
400×

100×

NONORGANIC

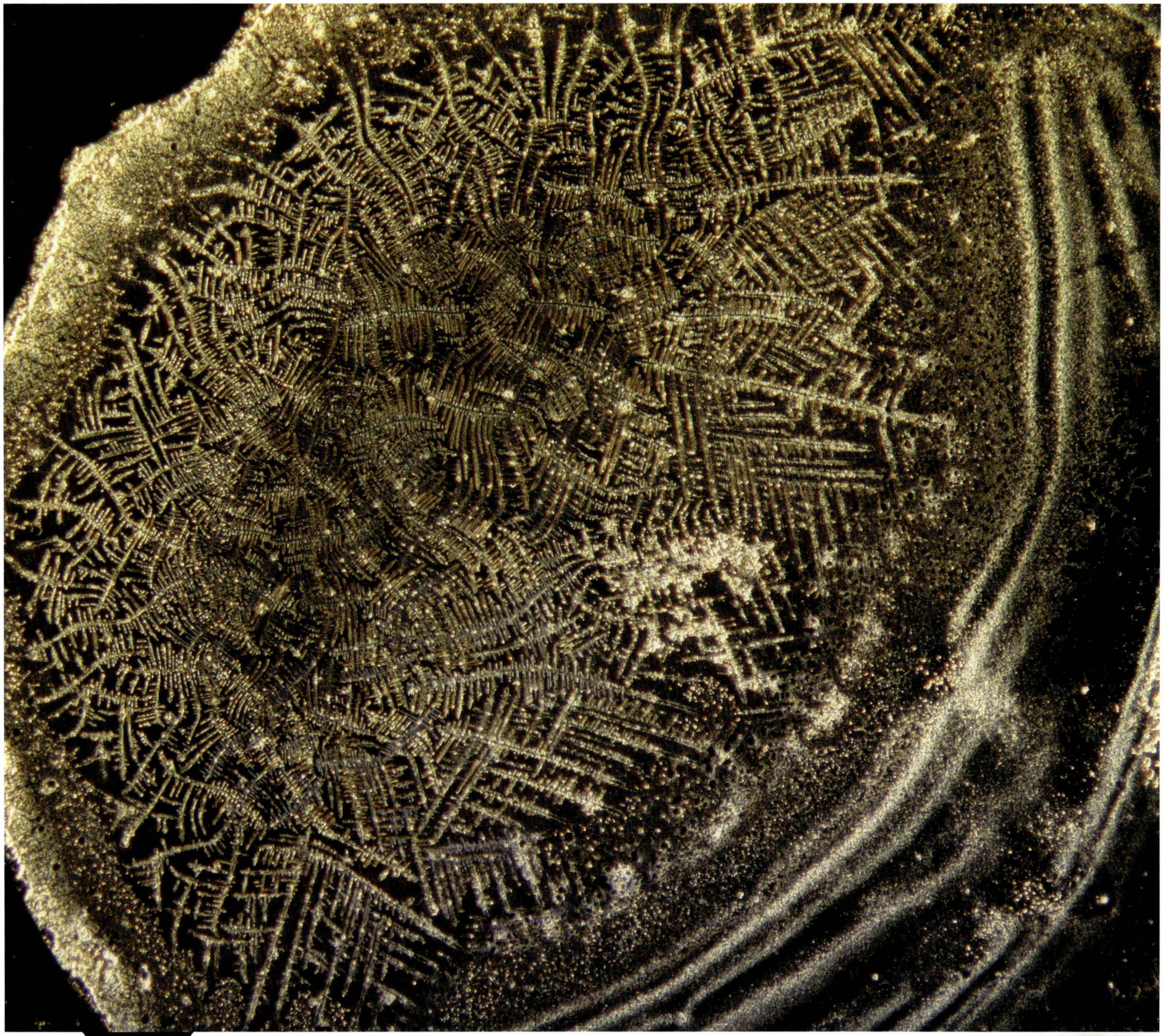

100×

ORGANIC

Carrots

The image shows only a few forms reminiscent of a little stardust, and connected rows of dots. The nonorganic carrots remember the life-energy that wanted to suffuse them, but their original forms and powers are lost.

The organic carrots, on the other hand, have captured life within a solar plexus, in which their secrets have been saved and coded into a characteristic carrot-labyrinth, clearly defined and always retrievable. When our cells find the key, they can nourish themselves with the mighty sun-wealth contained here.

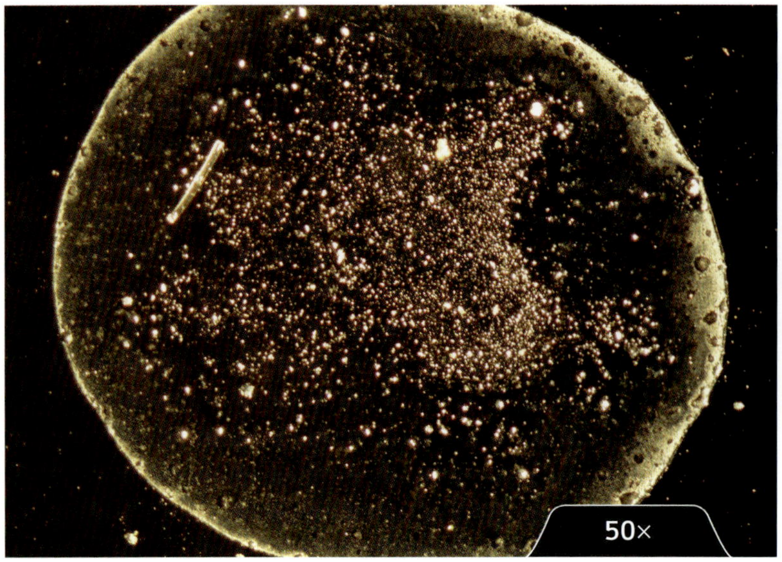

NONORGANIC

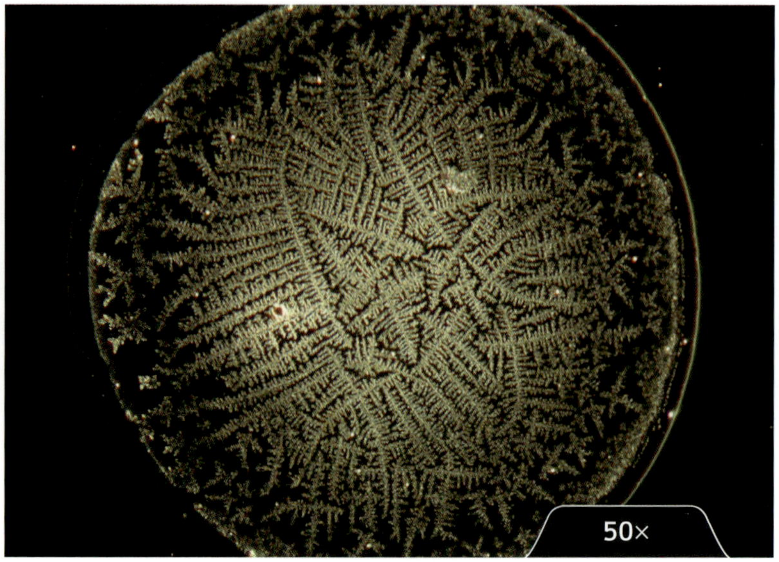

ORGANIC

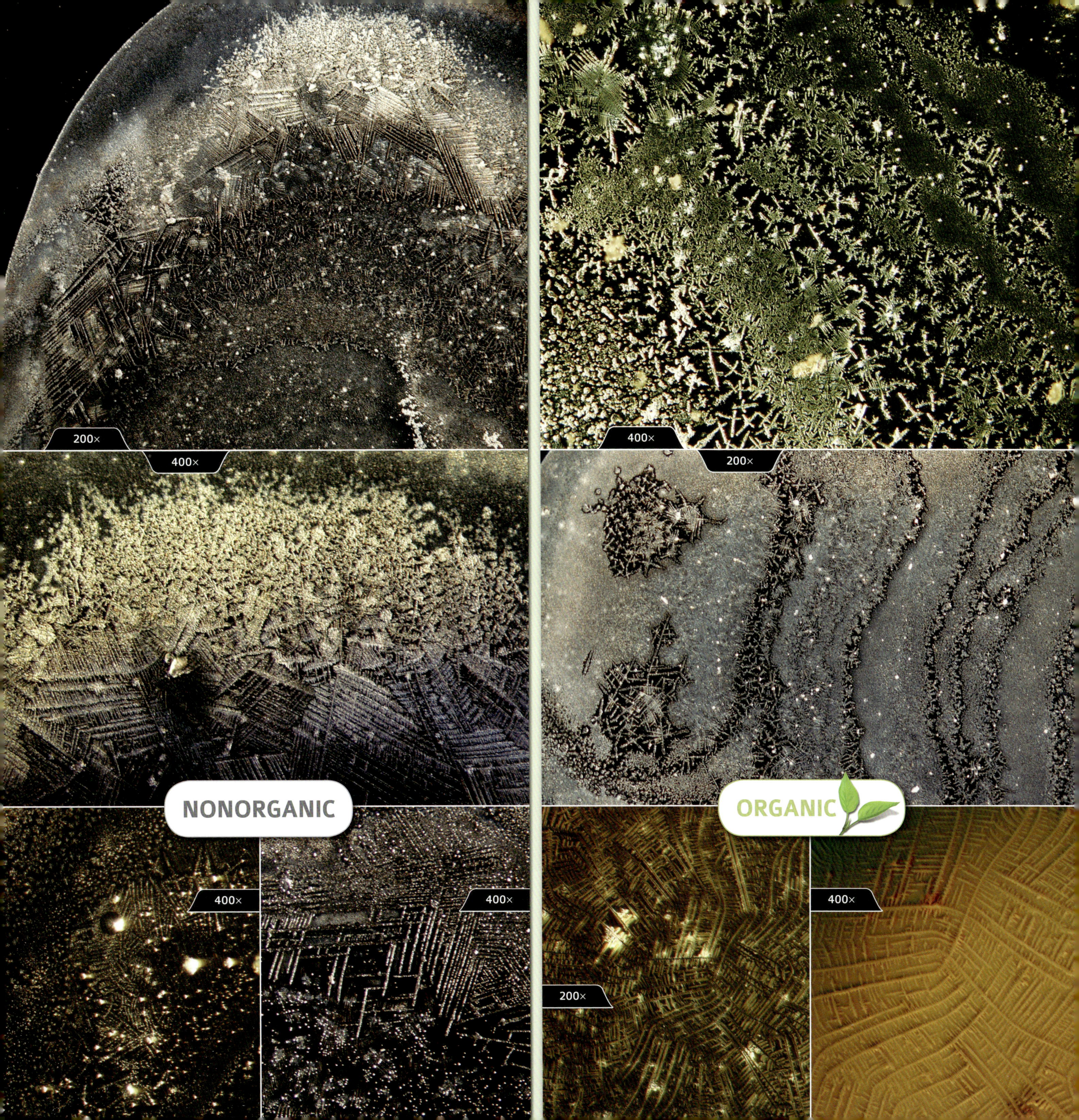
200×
400×
400×
200×
NONORGANIC
ORGANIC
400×
400×
200×
400×

200×

NONORGANIC

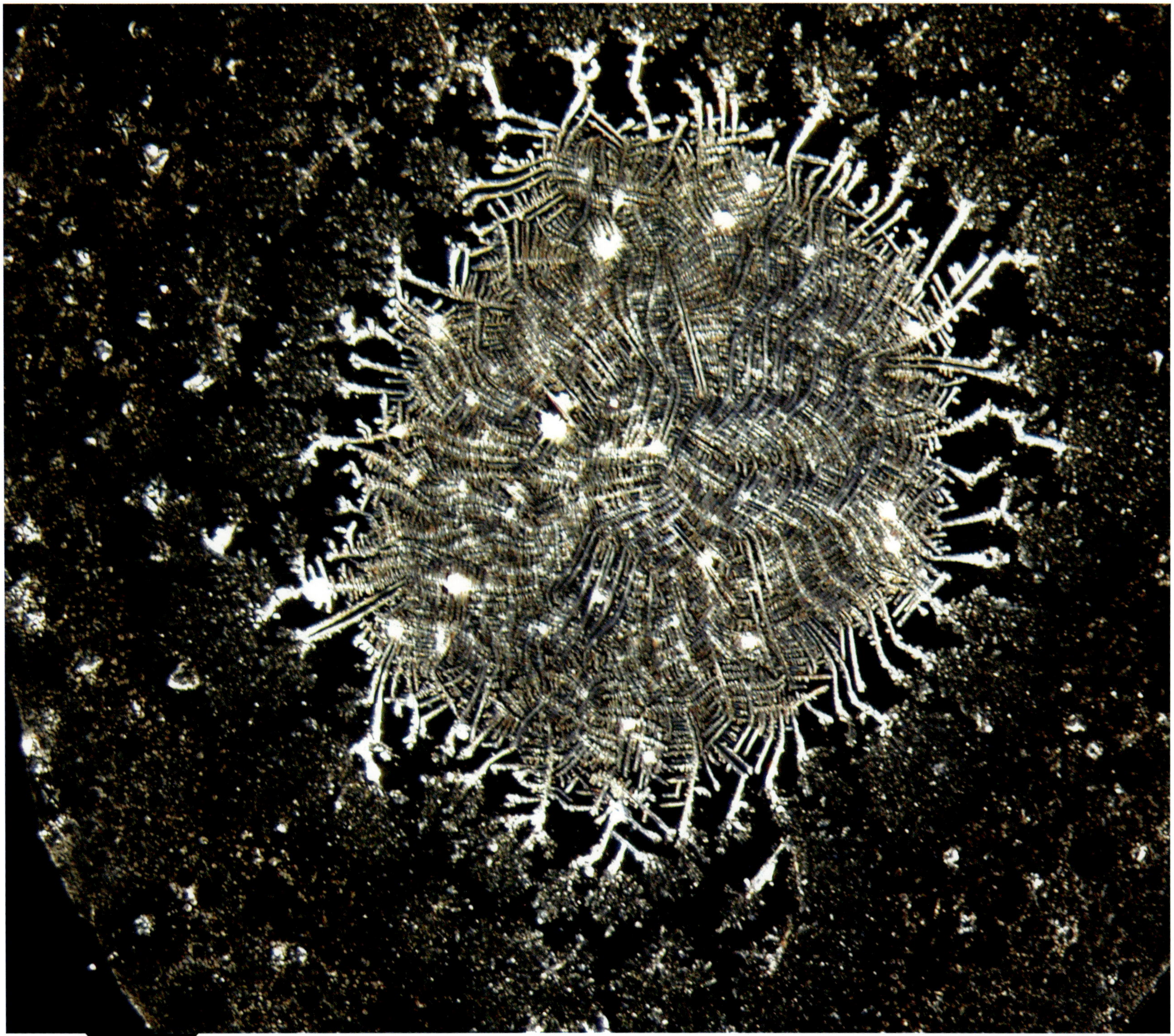

200×

ORGANIC

Potatoes

Those become strong who eat the power of potatoes. We can also see this power immediately in the nonorganic potatoes by the way it permeates the entire drop with plant-like shapes. Yet only in the organic potatoes do we observe the filigreed finesse in which this power flows, and the amazing beauty with which it crafts the most wonderful and artistically precious designs.

Looking once again at the nonorganic potatoes, we notice that the life-energy has gelatinized together like a starchy paste, whereas in the organic potatoes it has blossomed to its fullest beauty in the finest figures.

The common notion of the potato as a dull strength food is therefore only applicable to the nonorganic version, whereas our view of the nourishment that the potato can offer is illumined by the images of the organic potato, showing us food which offers the greatest strength in a most delicate unfolding.

NONORGANIC

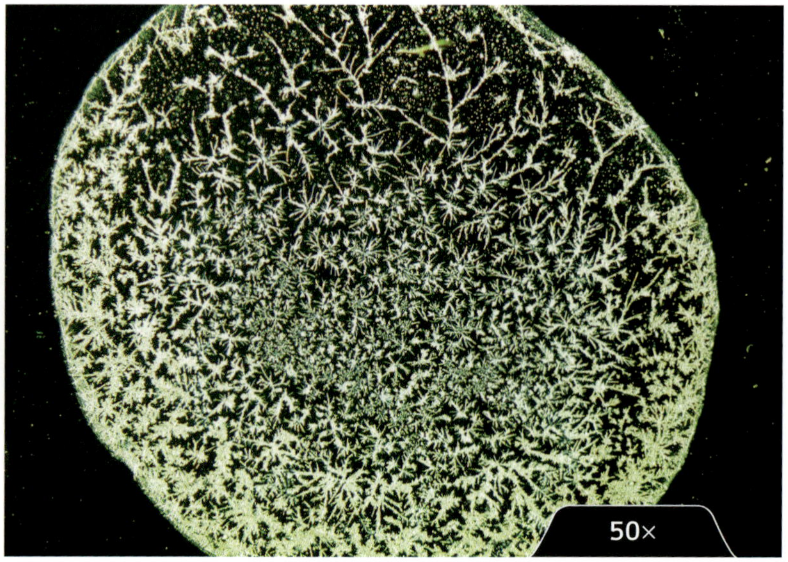

ORGANIC

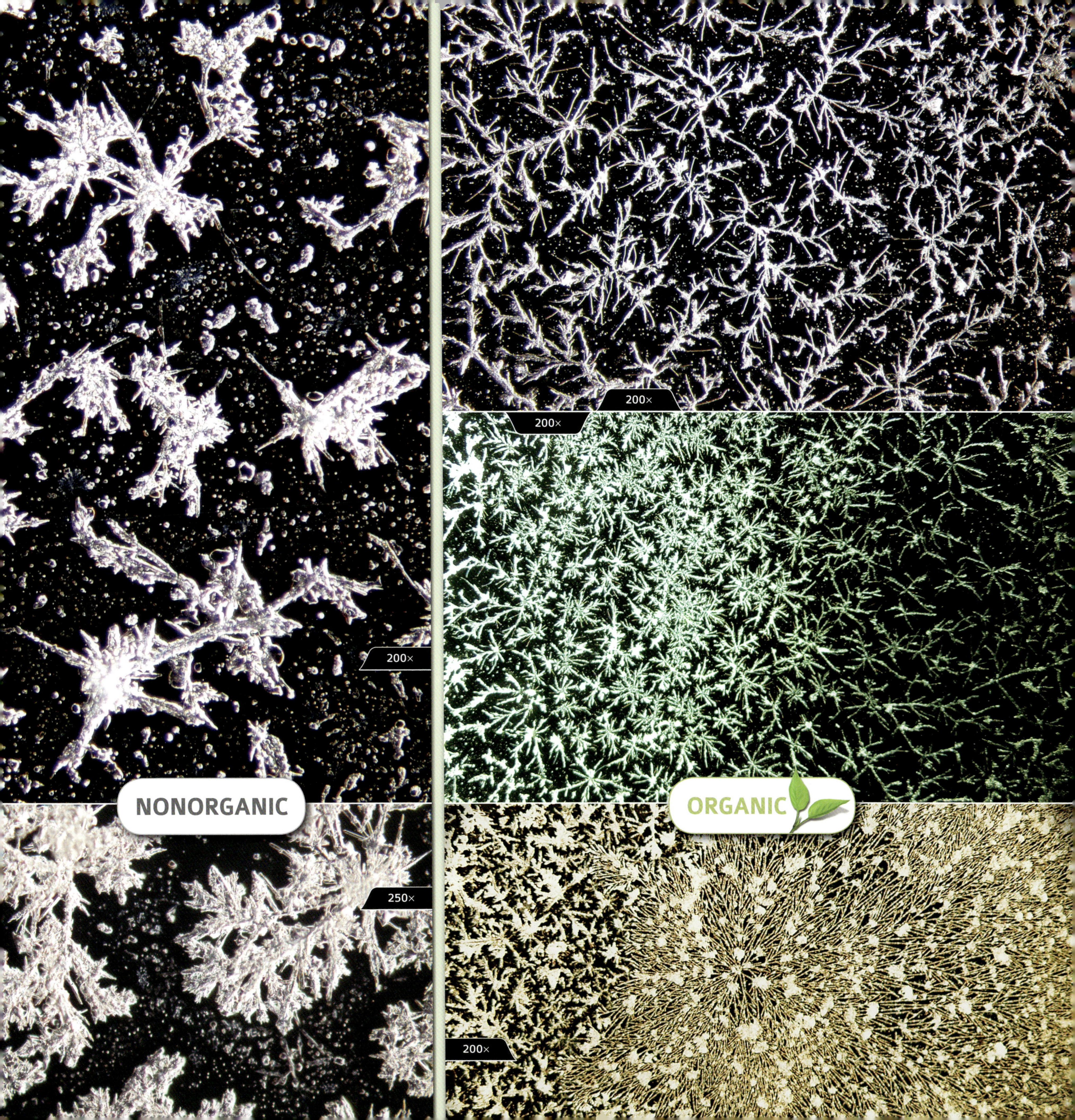
200×
200×
200×
250×
NONORGANIC
ORGANIC
200×

NONORGANIC

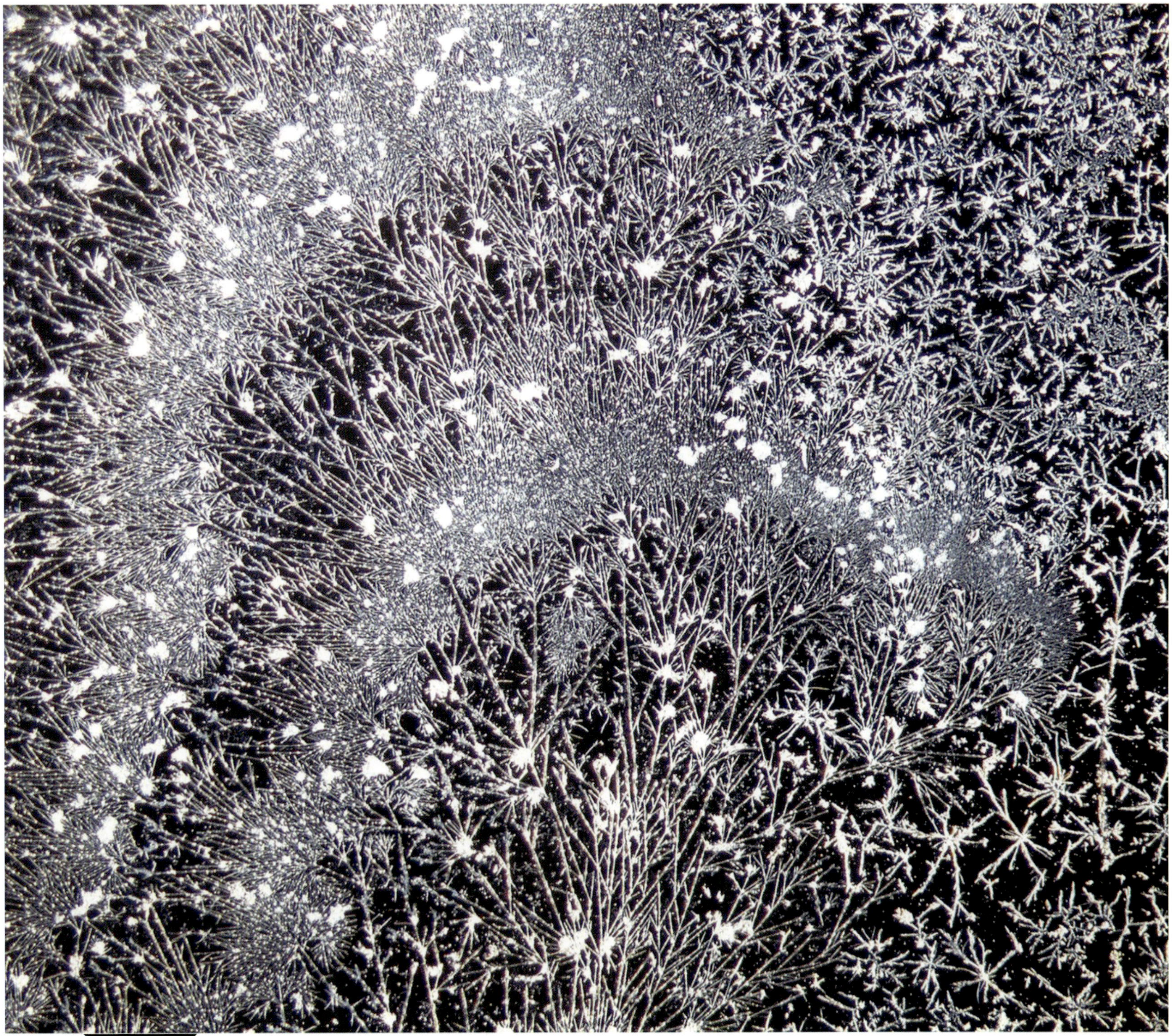

400×

ORGANIC

Tomatoes

The forms within the nonorganic tomatoes are relatively coarse and blurry. The parts that did not get squashed into rectangular grids have dissolved or disintegrated into an undefined pulp.

Within the organic tomatoes, networks of plant-like clusters immediately stand out, reminding us of honeycombs: the organic tomato uses the hexagonal honeycomb as a construction element on which it builds complete shapes, much like coral reefs that provide habitat. Out of the open hexagonal honeycombs, six-petalled blossoms are formed, indicating the "flower-of-life" principle discovered by the alchemists.

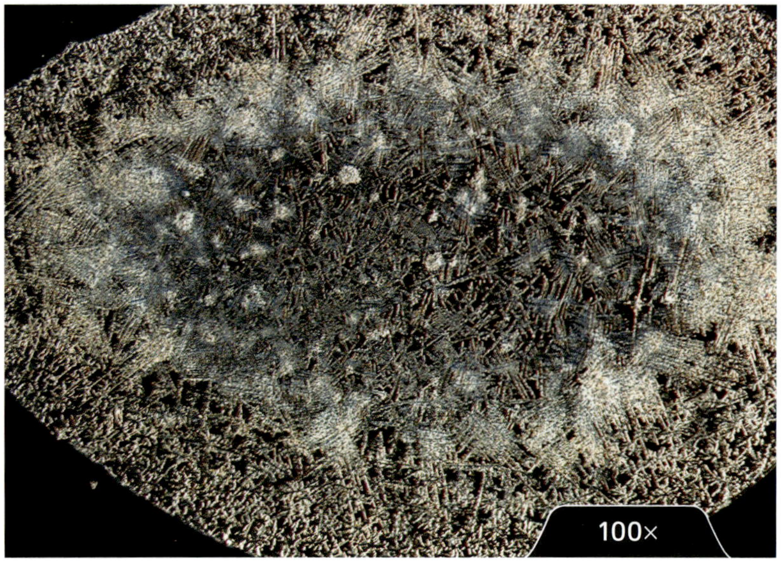

NONORGANIC

ORGANIC

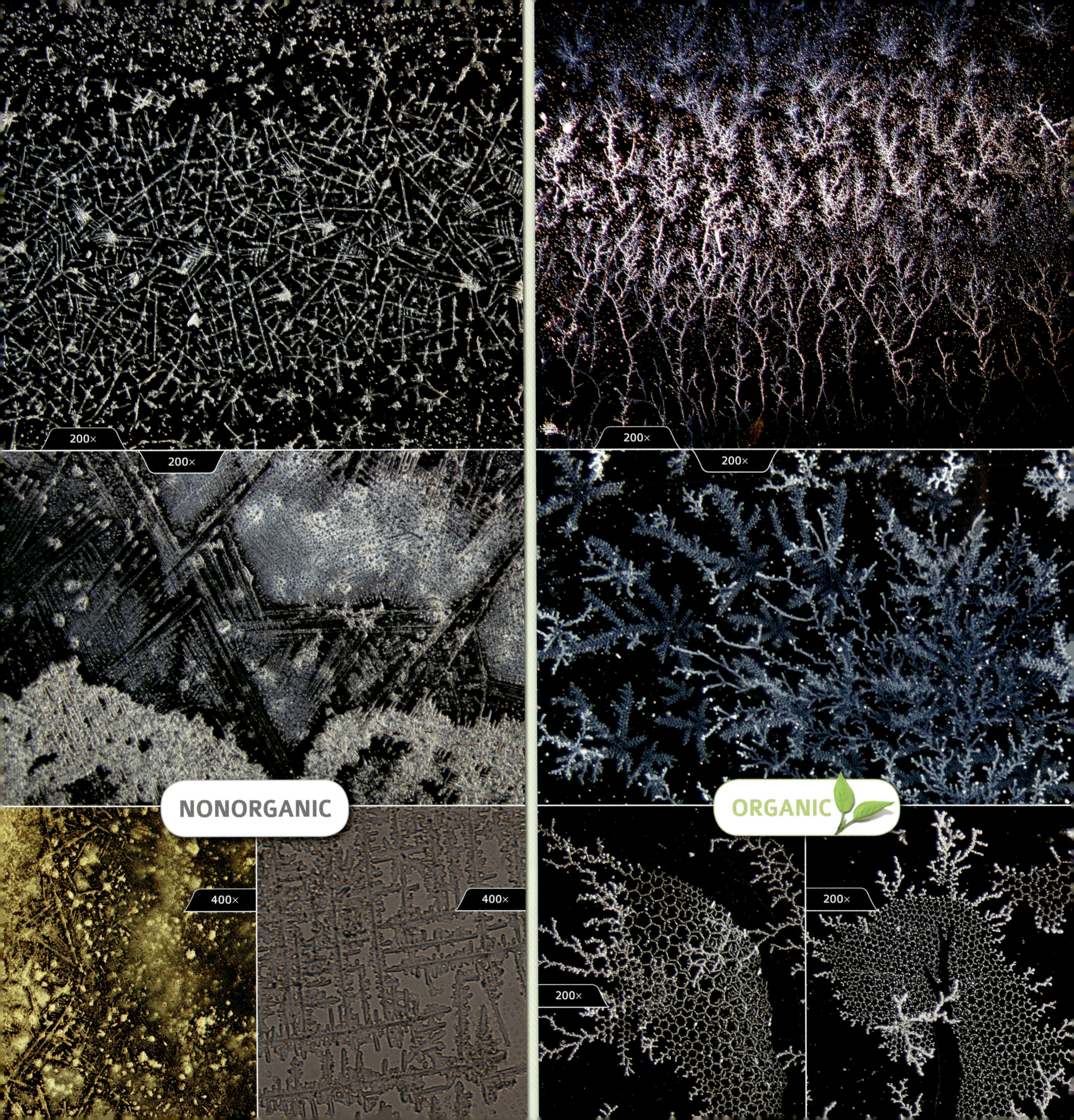
200×
200×
200×
200×
NONORGANIC
ORGANIC
400×
400×
200×
200×
200×

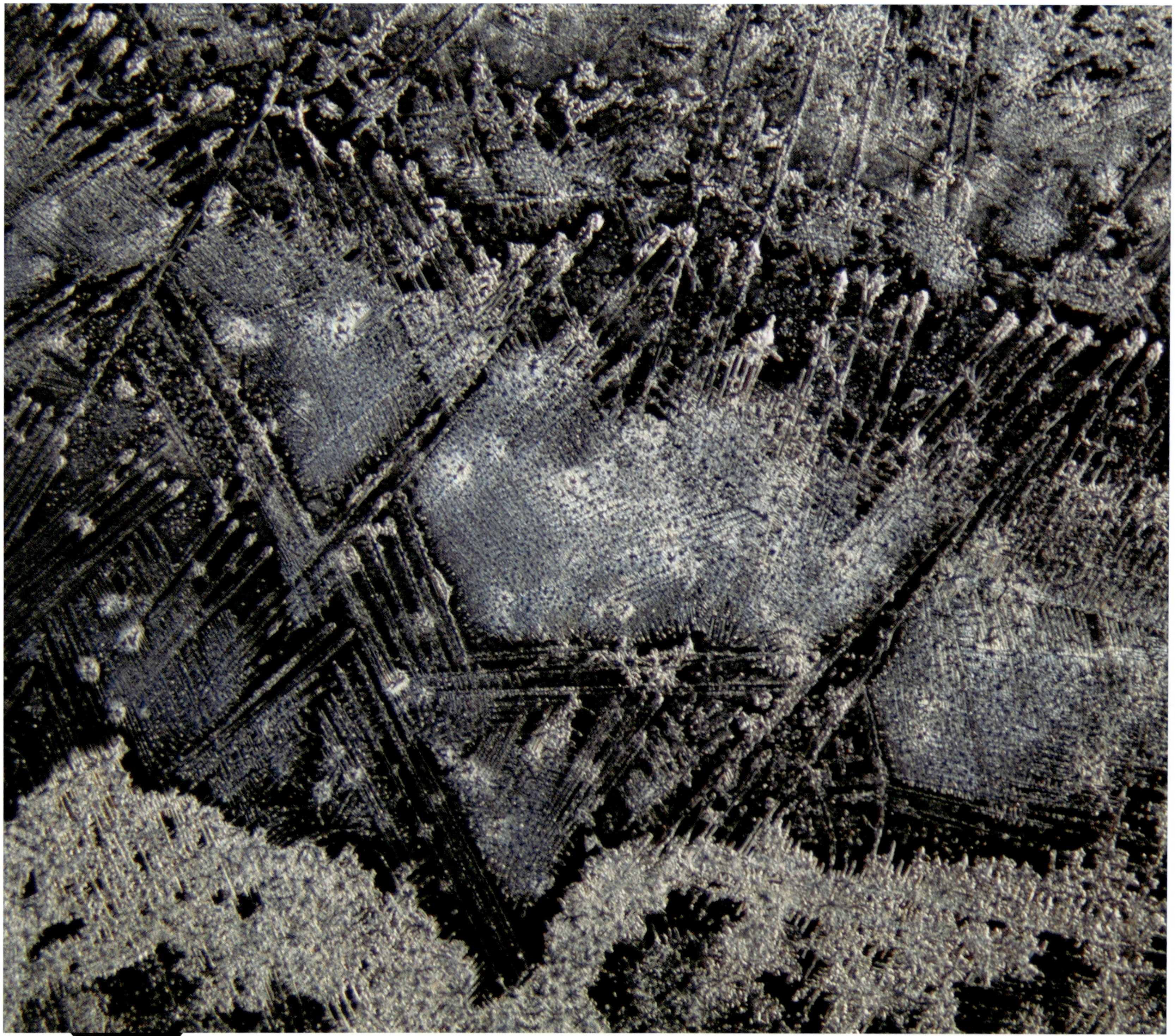

400×

NONORGANIC

400×

ORGANIC

Broccoli

While the nonorganic broccoli offers an image of lost information, confusion and destruction, the organic broccoli astonishes us with a phenomenal fullness, a sun-like center and a lavishly dynamic plant growth. It reminds me of the young boy in a well-known children's book who stands under giant trees in the jungle, feeling alive, sheltered and at home. Organic broccoli reveals tremendous order forces (life design principle), which in a plant-based diet has a balancing effect on our organism and perhaps can protect us from much harm. These images inspire us to use more organic broccoli. Of course it would be interesting to verify these results through clinical food tests on many different people. Maybe someone has already performed them, and we just don't know.

NONORGANIC

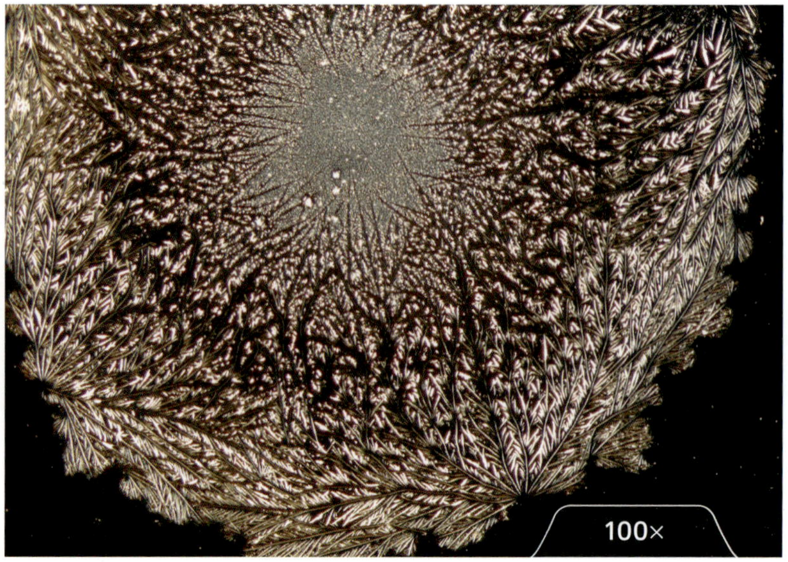

ORGANIC

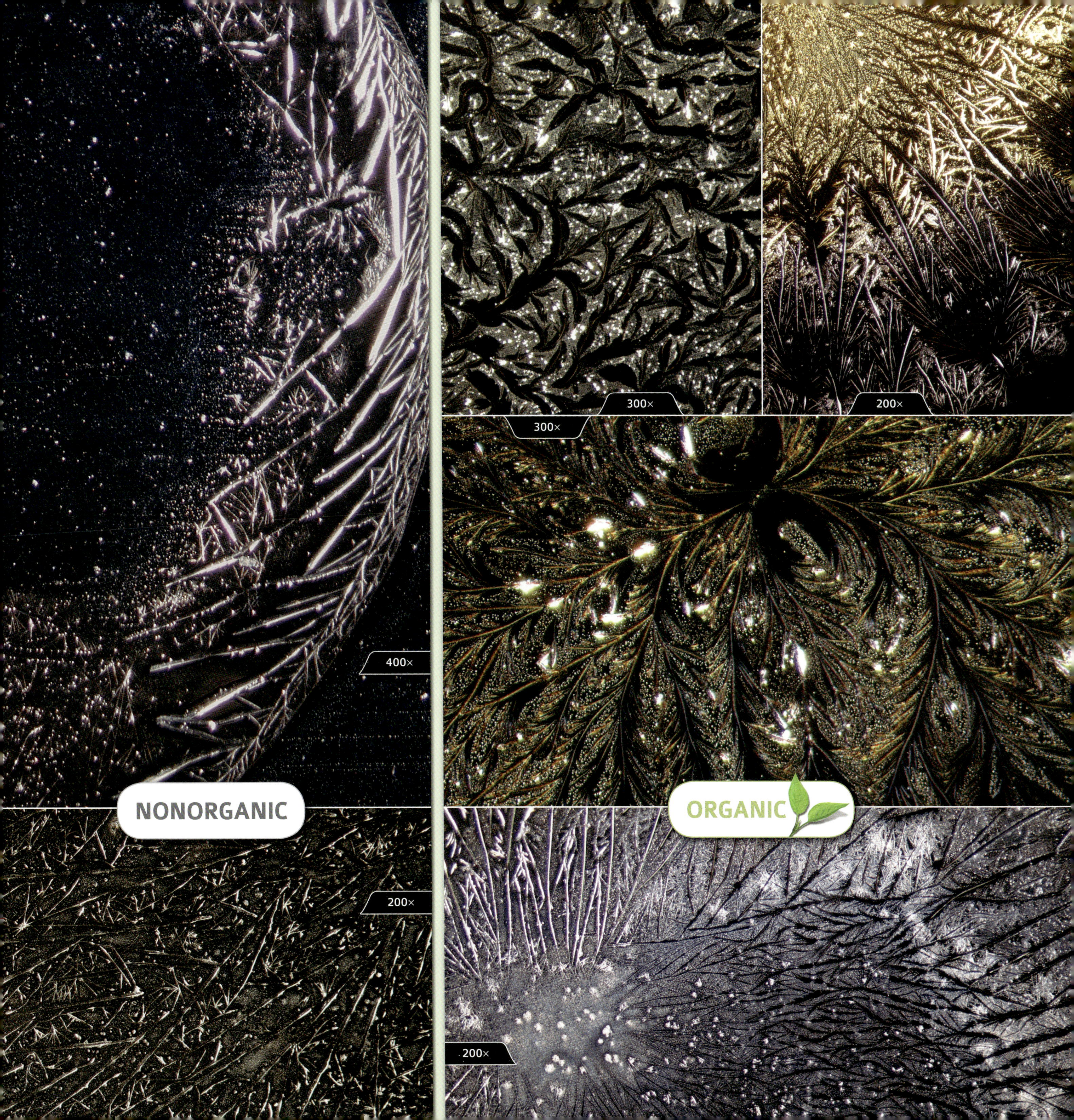

300×
200×
300×
400×
NONORGANIC
ORGANIC
200×
200×

400×

NONORGANIC

200×

ORGANIC

Avocado

The organic avocado could not show us its essence more clearly, vividly and even charmingly: the avocado *idea* penetrates deep into our visual understanding and is imprinted there. With its subtle reality it educates us with the essence, the *ideal imagination* of a natural avocado: this is what the life-energy of an avocado should look like.

Conversely, the nonorganic avocado is not only unable to form any kind of idea about an avocado, but with its information-holes, its stiffened forms and its lack of remembrance of its true nature, it deviates very strongly from the ideal. What kind of strange, thorny creature do we have here?

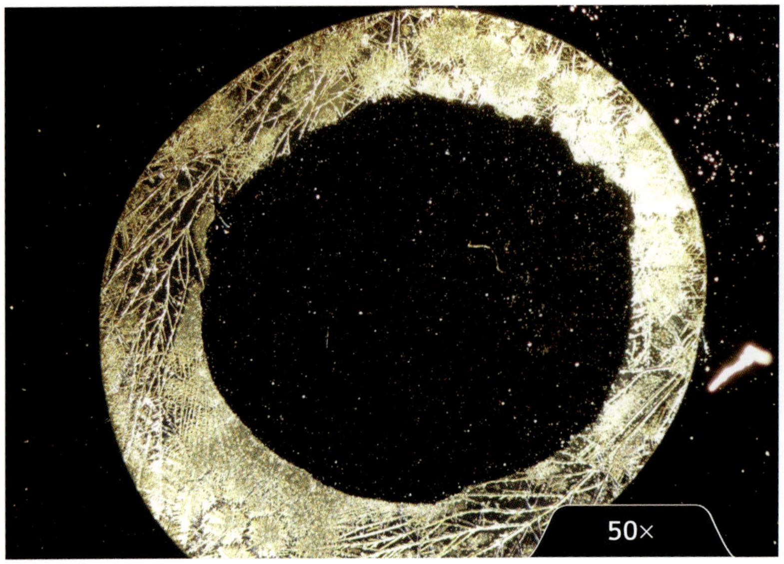

NONORGANIC

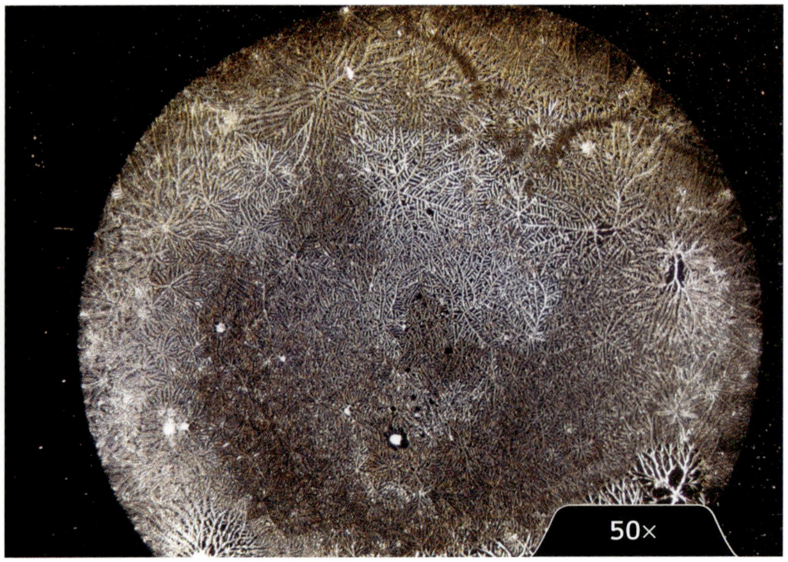

ORGANIC

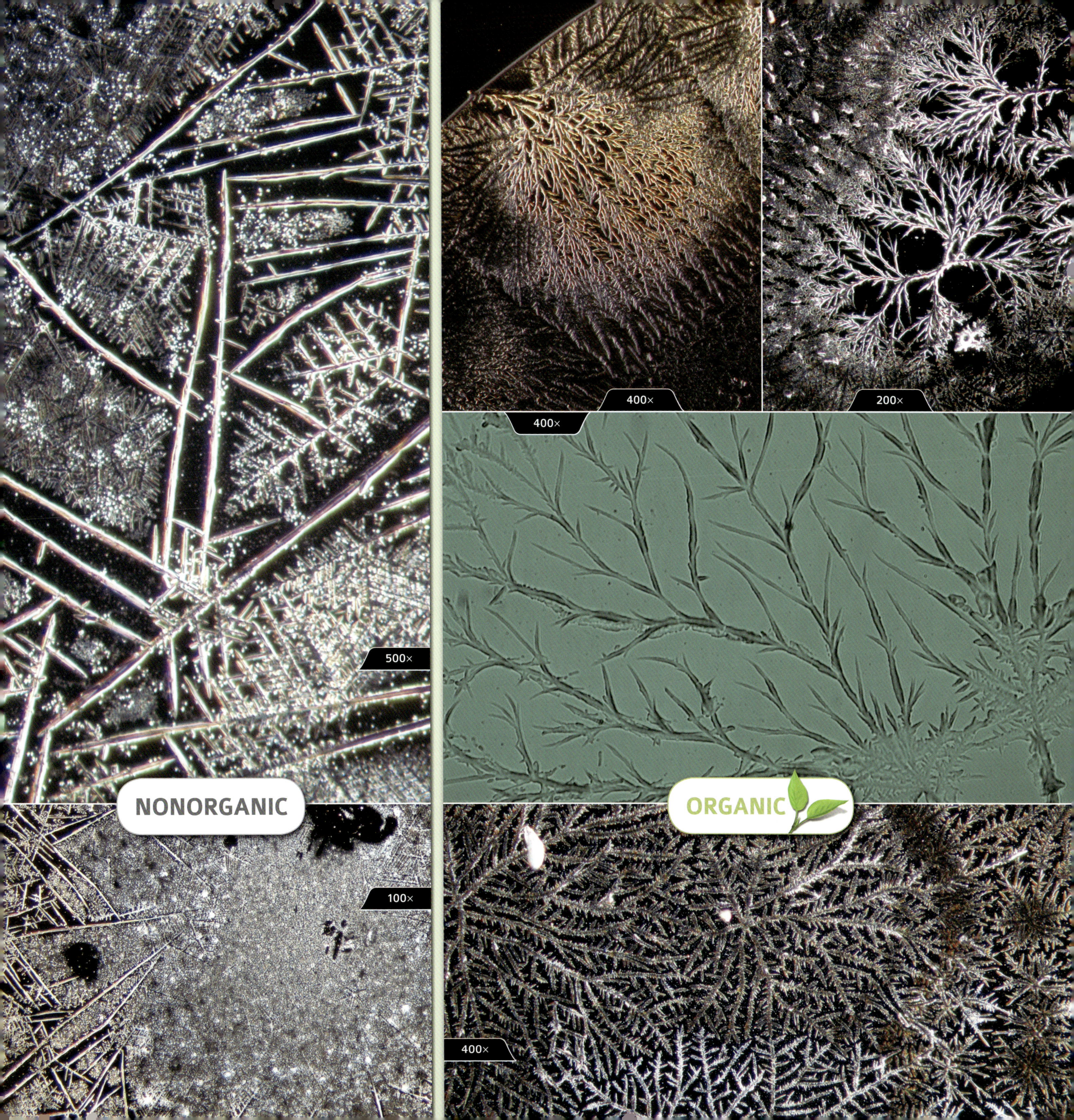

500×
NONORGANIC
100×
400×
200×
400×
ORGANIC
400×

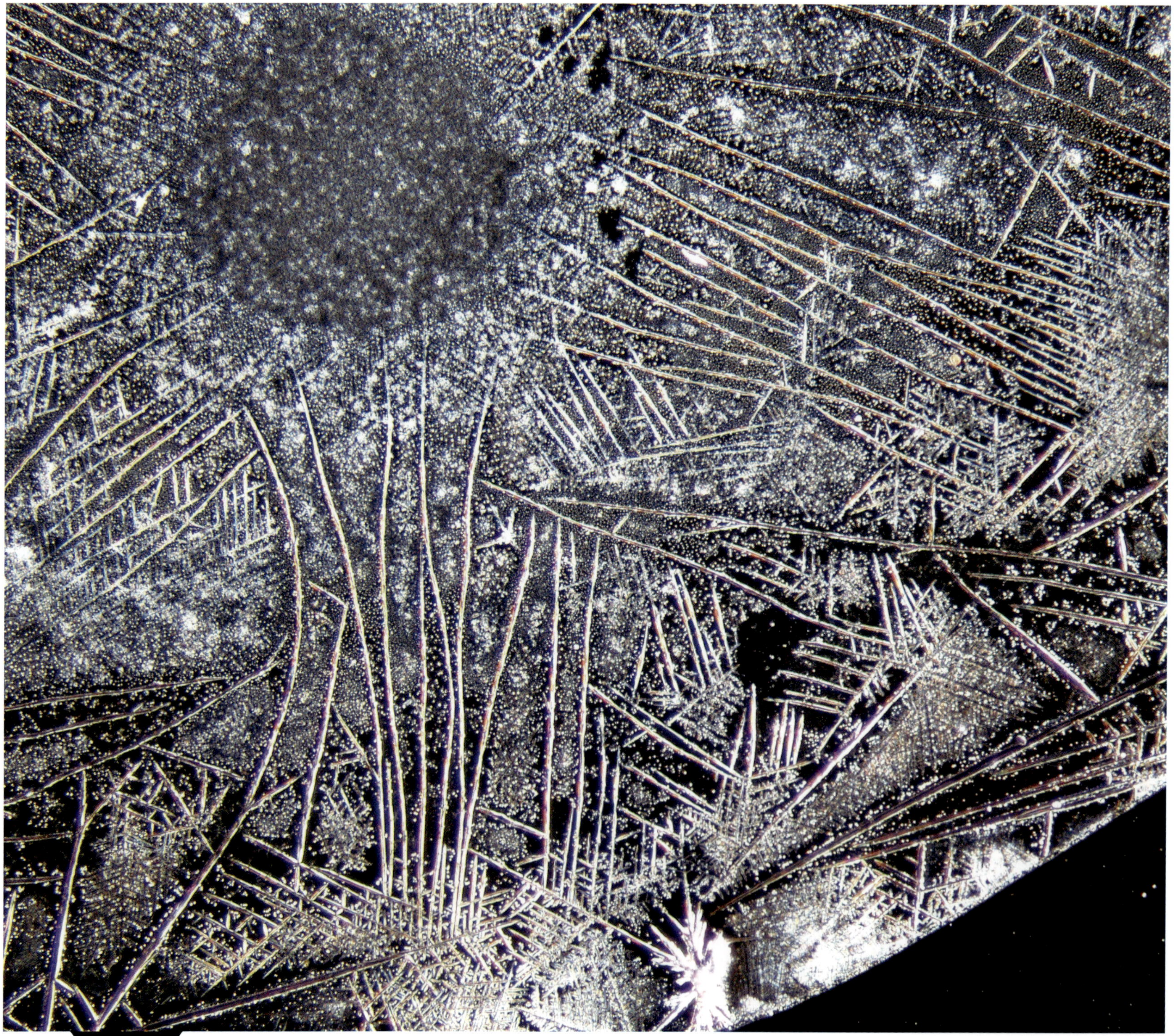

400×

NONORGANIC

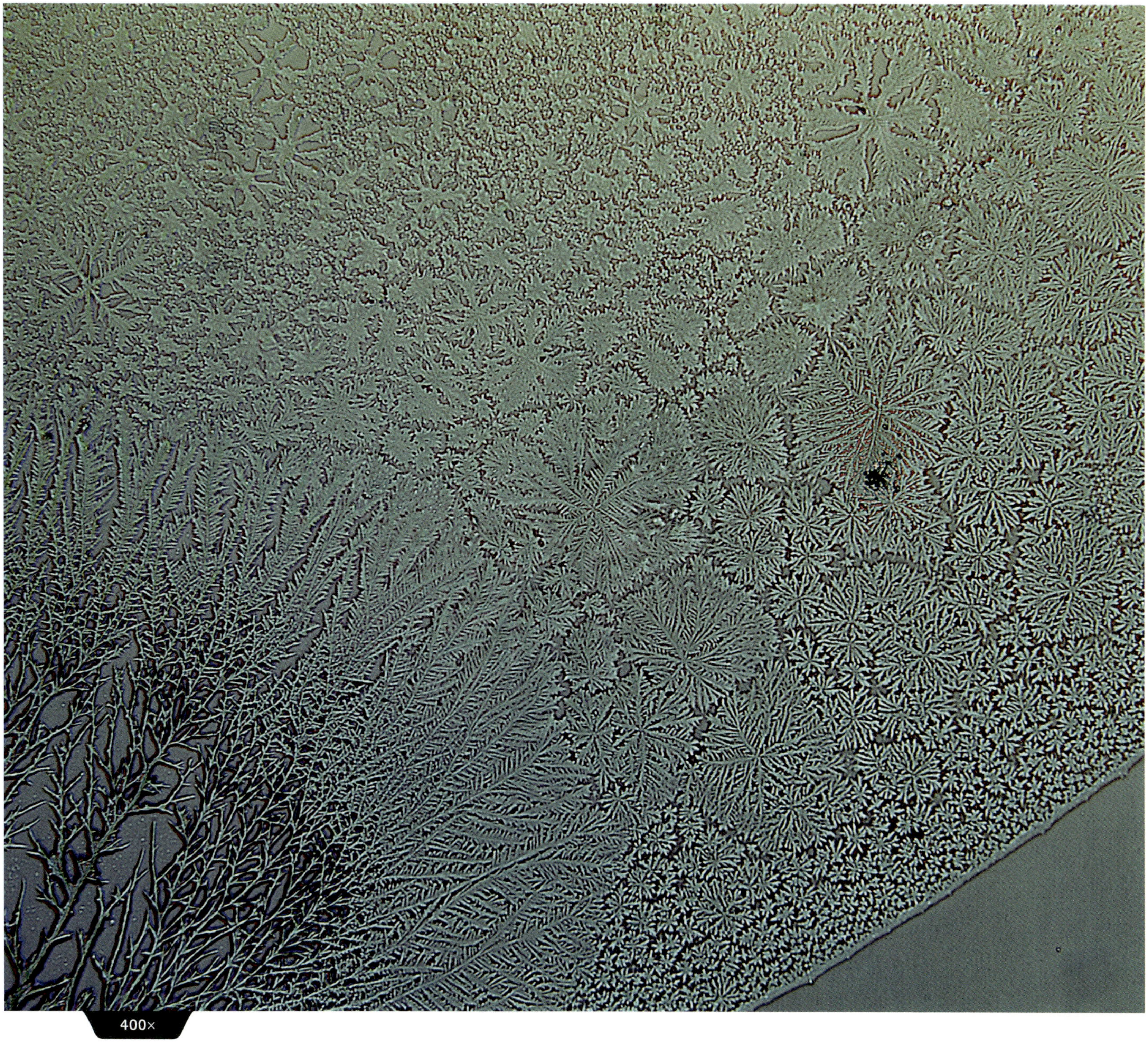

ORGANIC

Cucumber

From the watery character of the cucumber one would expect to see a strong outer structure with a weak interior, yet this is not the case. On the contrary, the organic cucumber reveals a rich abundance of powerful six-petalled plant-like forms, which proliferate at the edges into smaller shapes fading into invisibility. Only a color filter of the microscope can make them visible.

The nonorganic cucumber cannot maintain this design: where the plant-like forms should be burgeoning, a great emptiness yawns. The remaining form-power is pushed to the edges and stored up like sand in blurry sedimentary layers. The expanding abundance of powerful plant-like forms is no longer present.

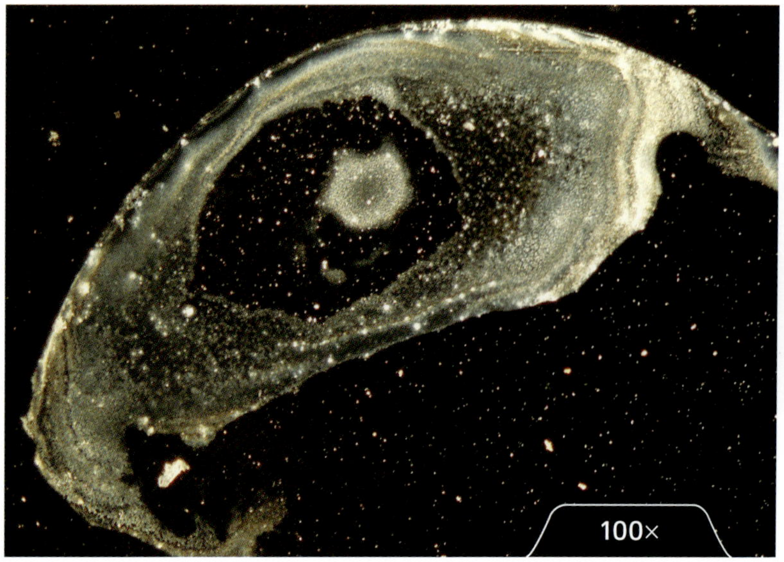

NONORGANIC

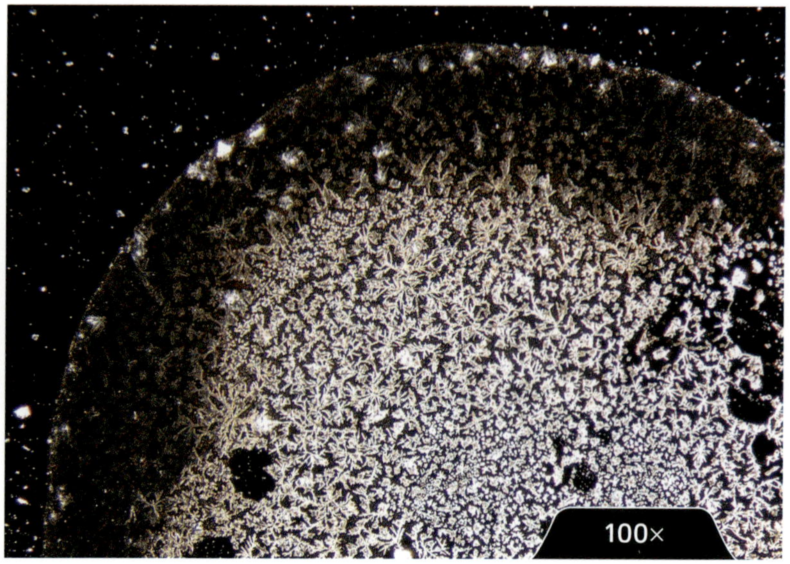

ORGANIC

NONORGANIC

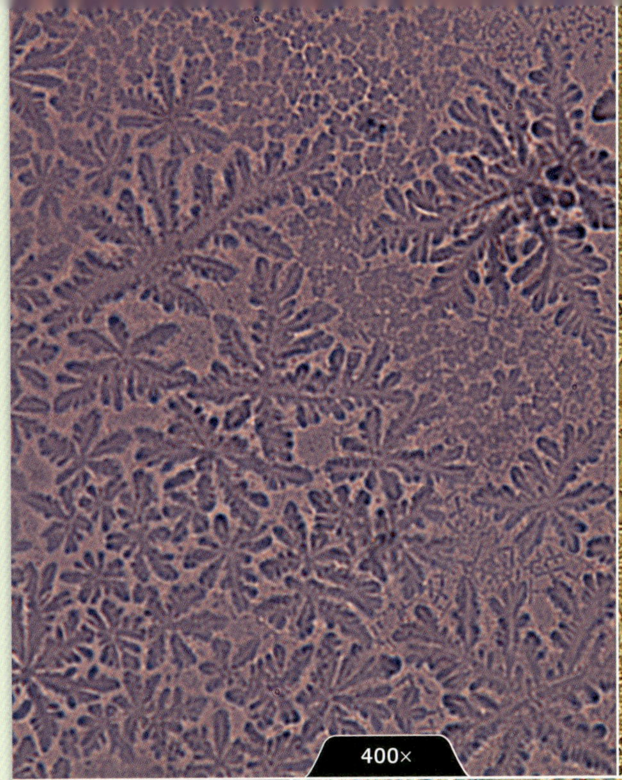

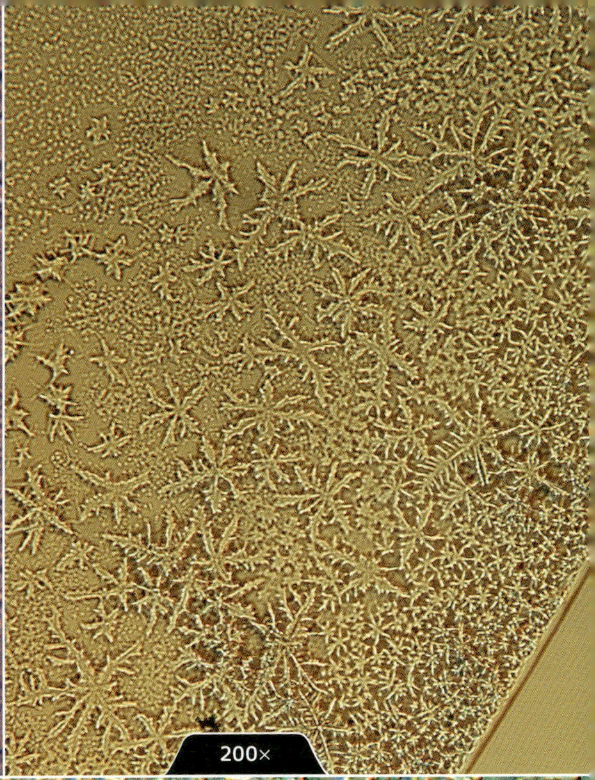

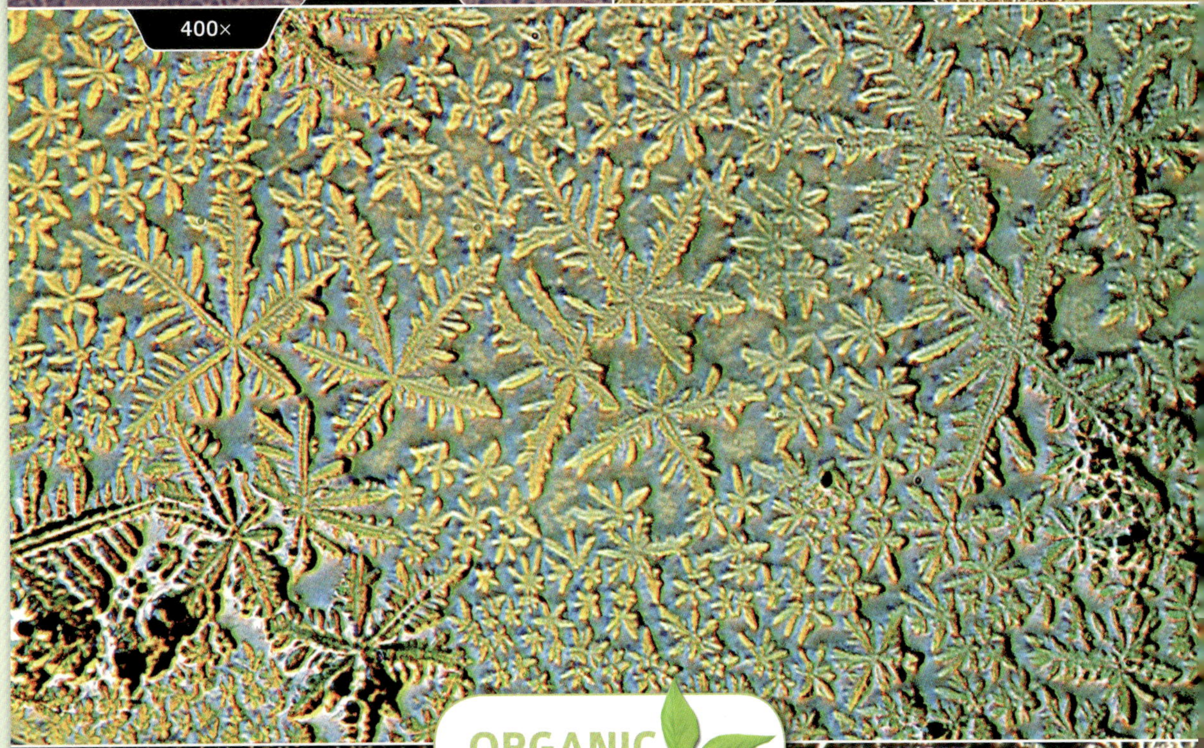

ORGANIC

NONORGANIC

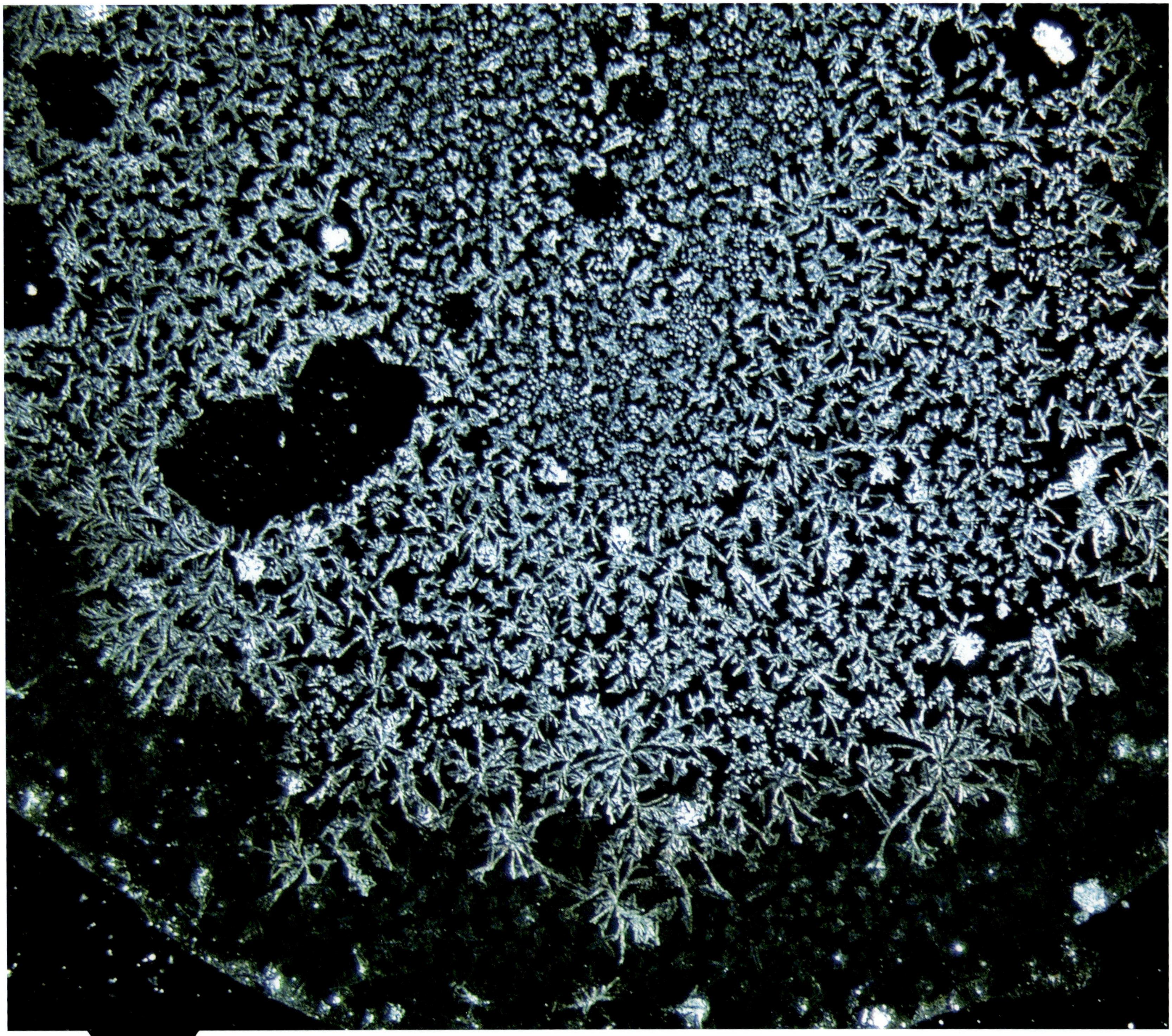

200×

ORGANIC

Cabbage

The information of both the organic and the nonorganic cabbage fill up the entire droplet. Cabbage must have an enormously strong character which does not allow itself to be affected by artificial fertilizers and agricultural chemicals, one would think at a first glance. But while in the organic cabbage a fabric of fine plant-like forms develops in the space between the strong channels, the nonorganic cabbage is unable to offer the same information. Even when the image hints at a well-developed structure, the enlargement reveals merely dotted lines; the finer structures have collapsed as if destroyed by fire.

NONORGANIC

ORGANIC

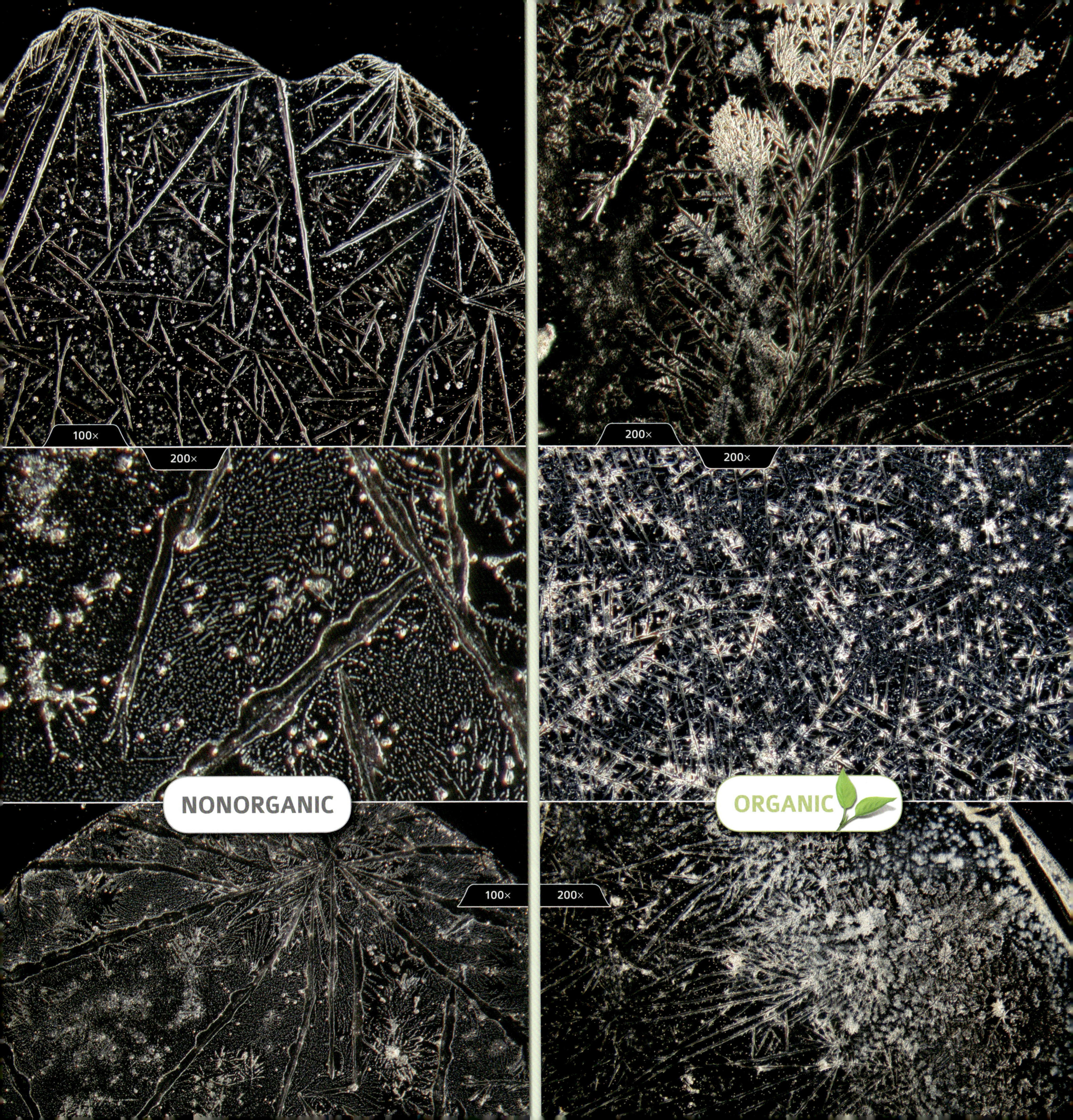
100×
200×
200×
200×
NONORGANIC
ORGANIC
100×
200×

200×

NONORGANIC

200×

ORGANIC

Spinach

Spinach has a very strong and gracefully built order force (life design principle), but it shows itself only in the organic spinach, where it forms a rich multiplicity with many six-petalled plant-like forms. In the nonorganic spinach the fine structure has withered away to rubble, whose fragments fail to show any traces of life even when enlarged 500 times. Still, some information does remain inside the droplet.

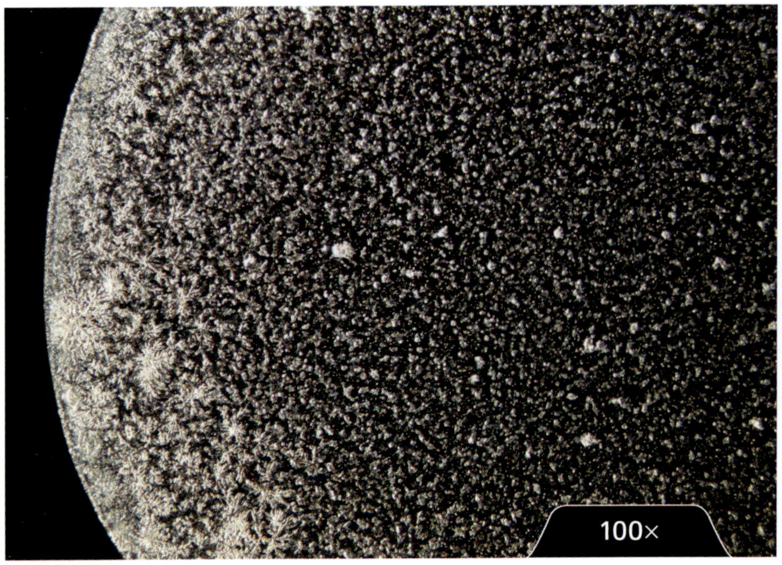

NONORGANIC

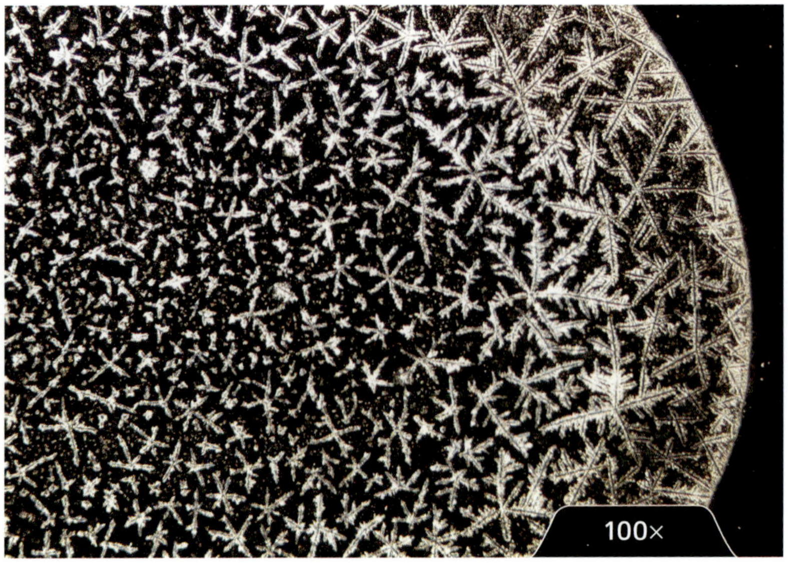

ORGANIC

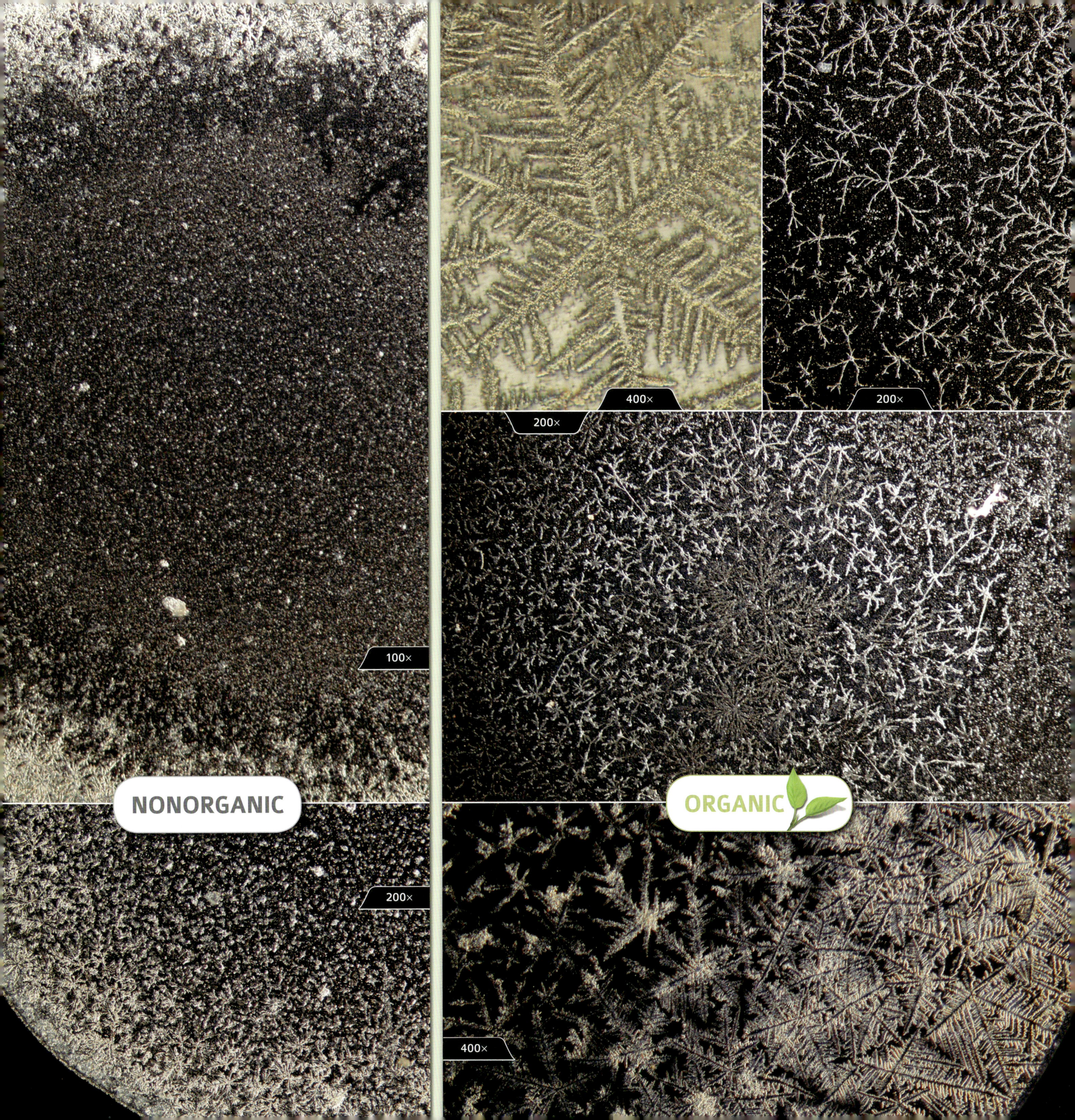

400×
200×
200×
100×
NONORGANIC
ORGANIC
200×
400×

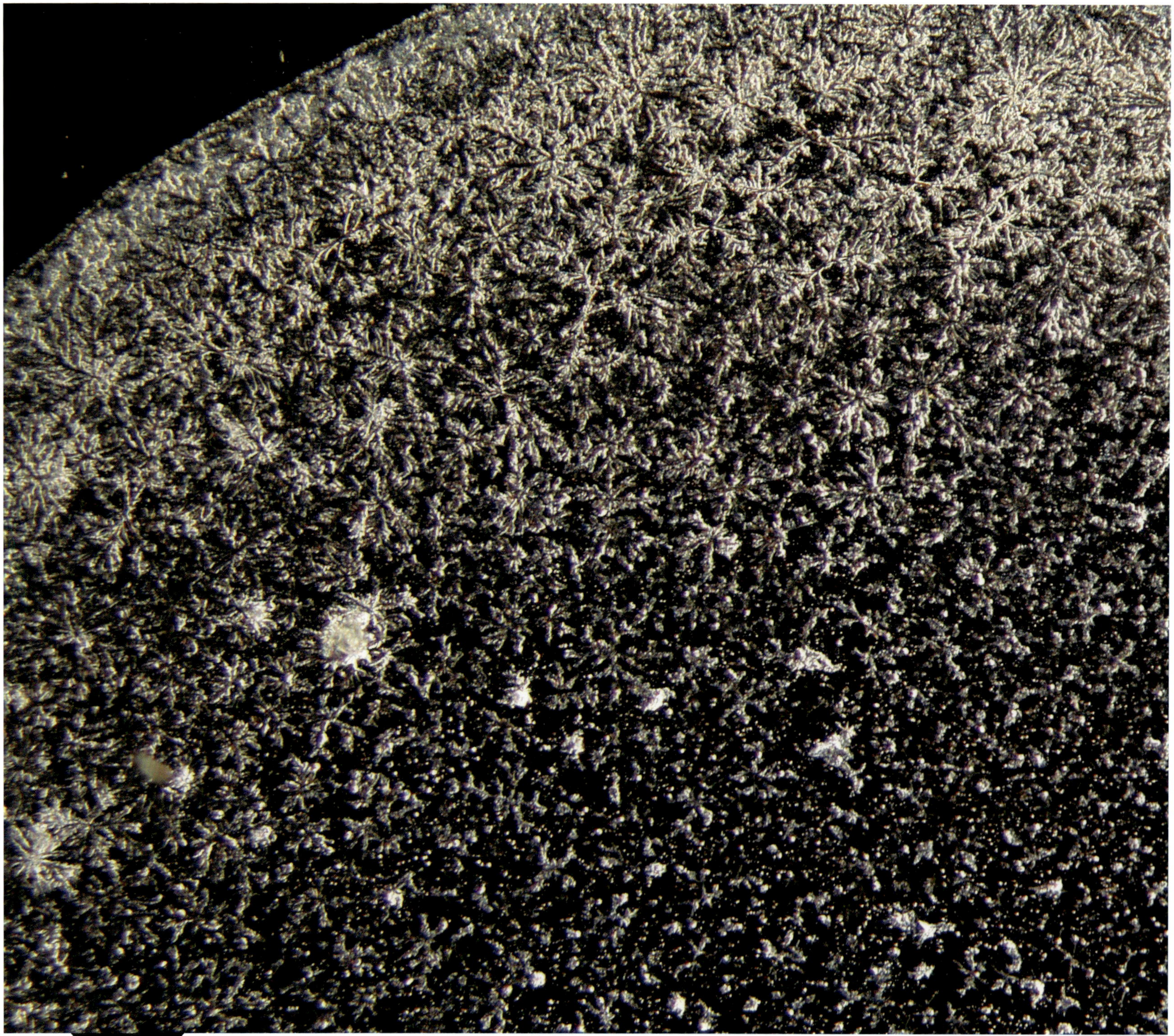

200×

NONORGANIC

200×

ORGANIC

Eggplant

The organic eggplant is permeated by a subtle plant-like fabric whose texture is stronger in the middle and finer around the edges. Life and beauty play together harmoniously.

The nonorganic eggplant would like to follow its natural form, but cannot. The end results are big holes and largely untextured debris with a few last, coarsely shaped plant-like forms. The connections are cut off, the graceful design is lost, the beauty forgotten.

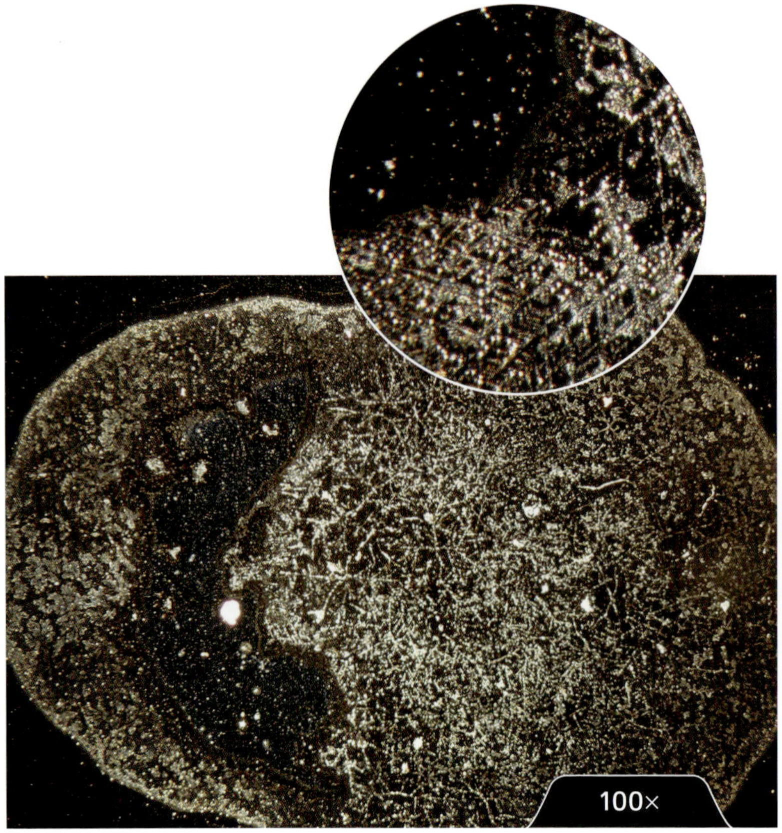

NONORGANIC

ORGANIC

400×
200×
NONORGANIC
200×
200×
400×
400×
ORGANIC
400×
200×

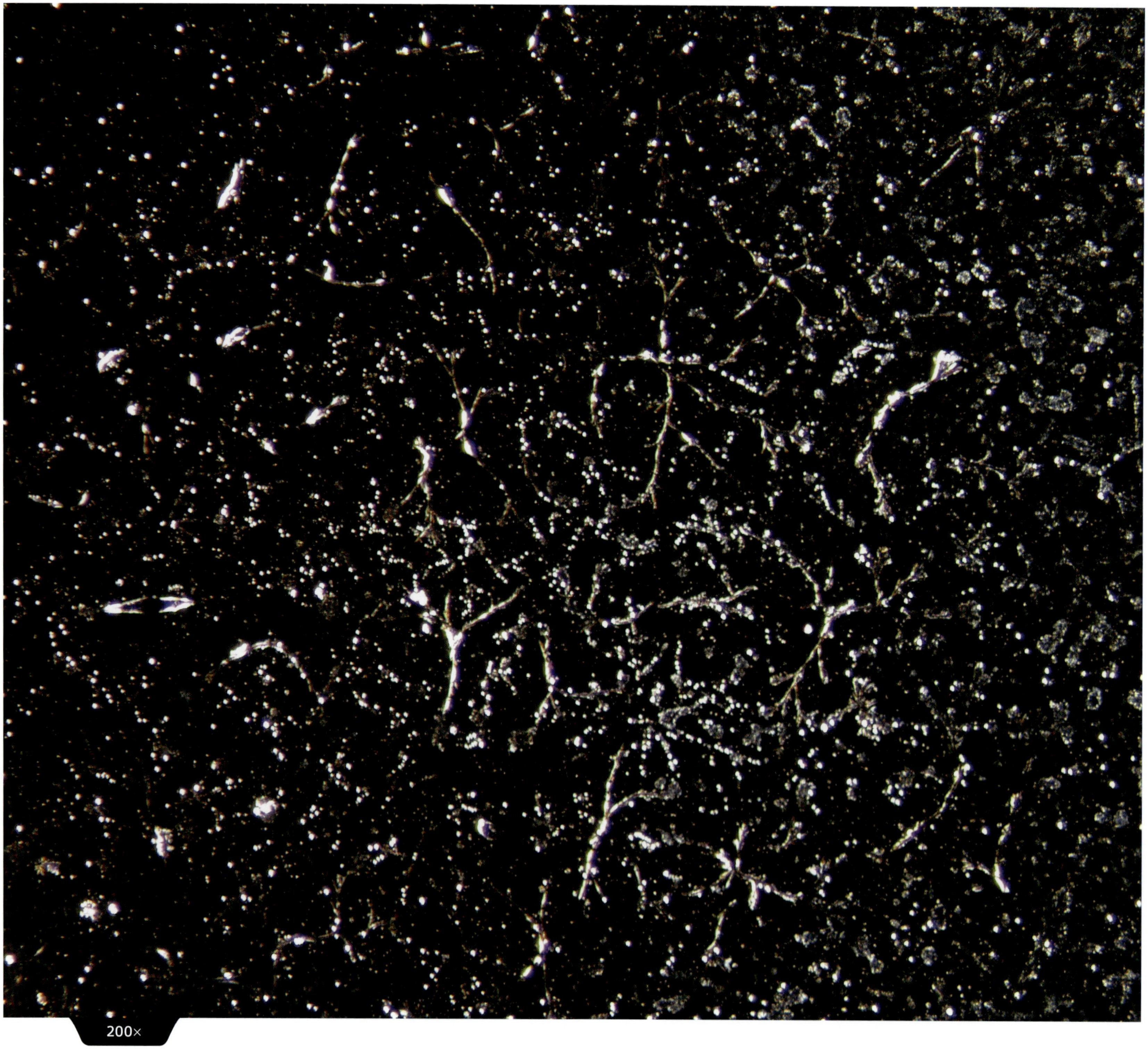

NONORGANIC

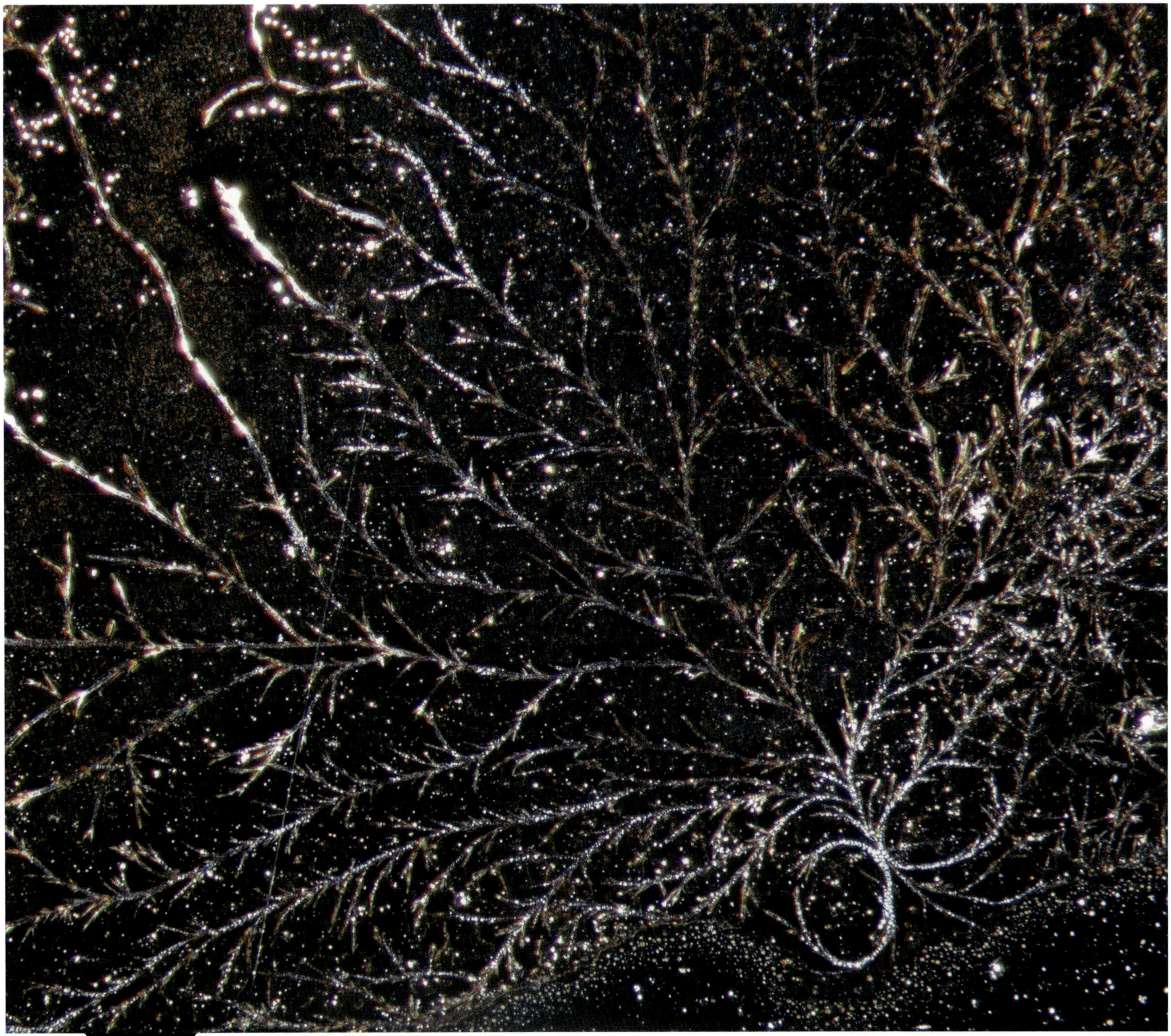

200×

ORGANIC

Zucchini

Dense foliage fills the organic zucchini with an abundance of six-petalled plant-like forms, a wonderful vitality that meets the observer with many "flowers-of-life". In the nonorganic zucchini an accident has happened: instead of "flowers-of-life" it offers an image of destruction. It cannot sustain any angular structures at all, and shows a gaping empty space.

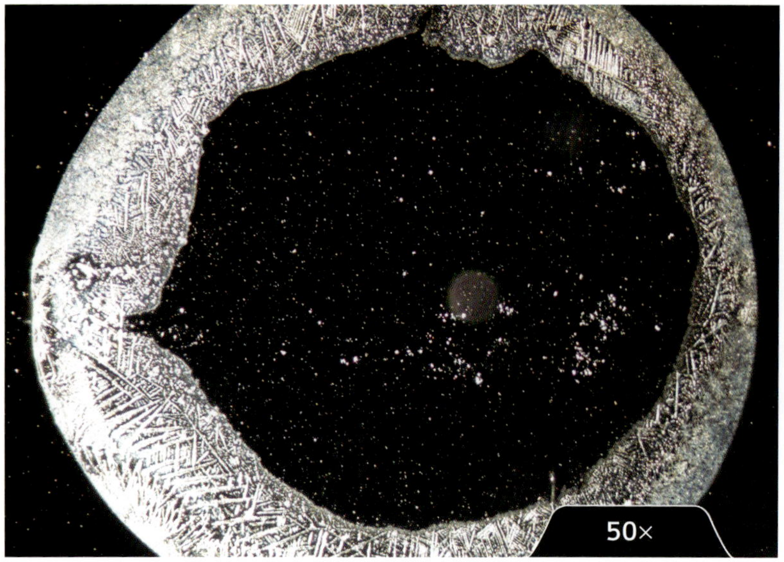

NONORGANIC

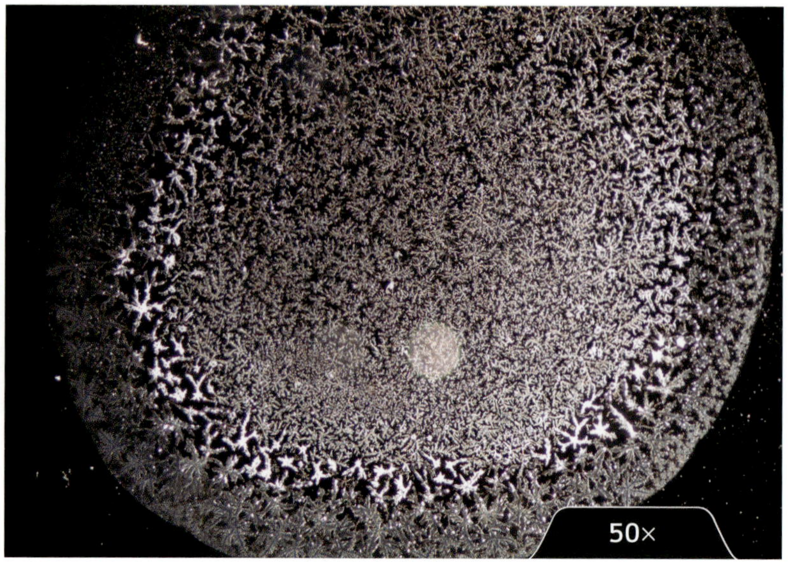

ORGANIC

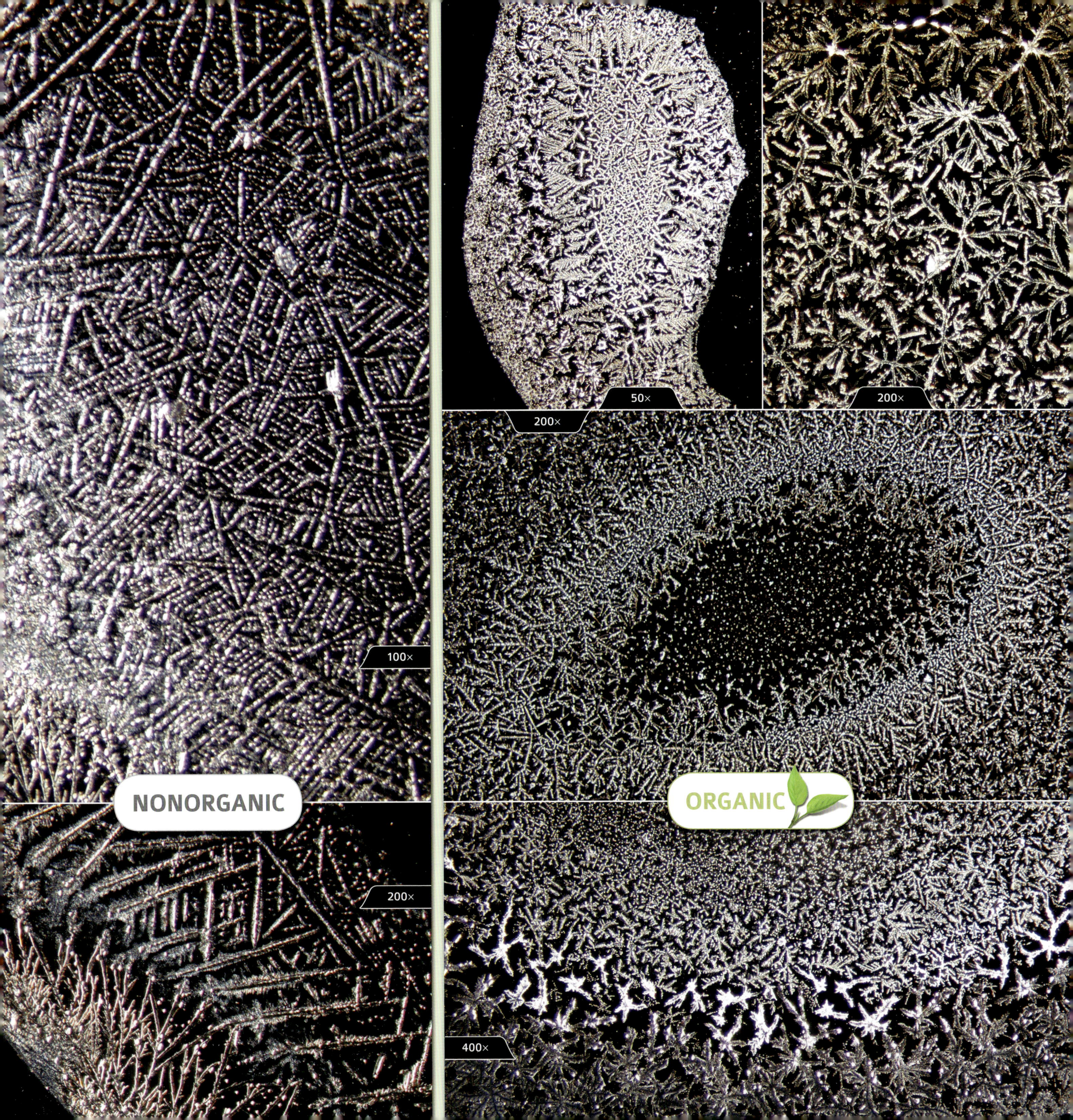

100×
NONORGANIC
200×
50×
200×
200×
ORGANIC
400×

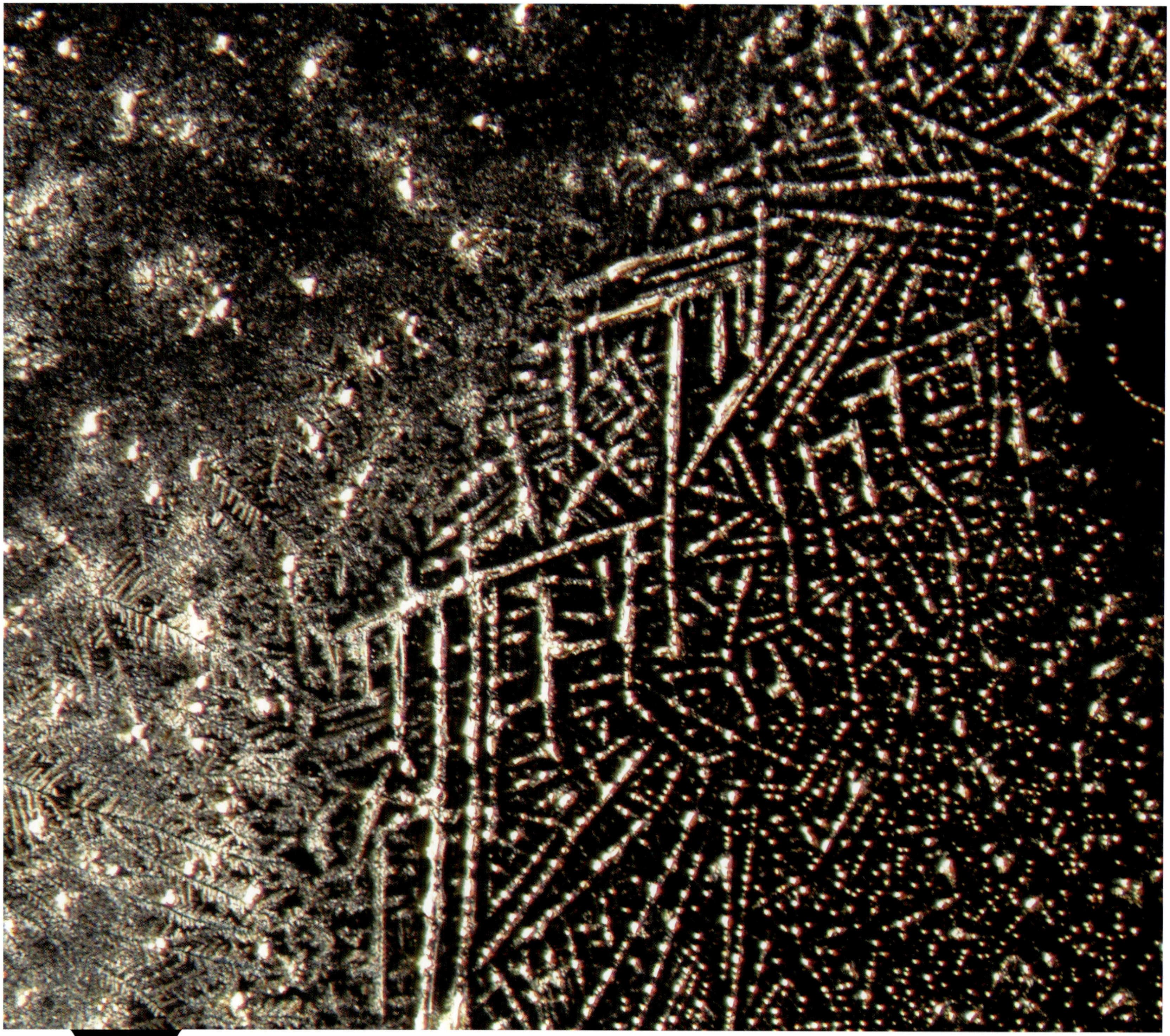

200×

NONORGANIC

250×

ORGANIC

Celery

In the organic celery two forms fill the space on the inside of the droplet: one looks like a garden of flower buds, the other resembles a jungle of roots. Both are plant-like, organic manifestations of an extremely vibrant order force. The nonorganic celery is unable to call up this living information. This results in an inner wasteland, on whose borders a bulwark of plant-like memories is reared up, almost like defense works against the advancing desert. These forms are characterized by a struggle against something menacing. They differ vastly from the organic celery, where a harmonious sprouting shapes a hospitable world.

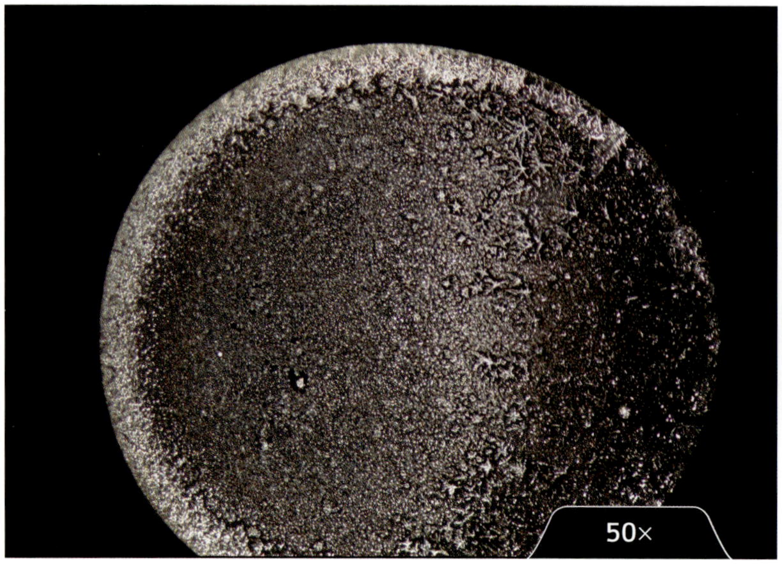

NONORGANIC

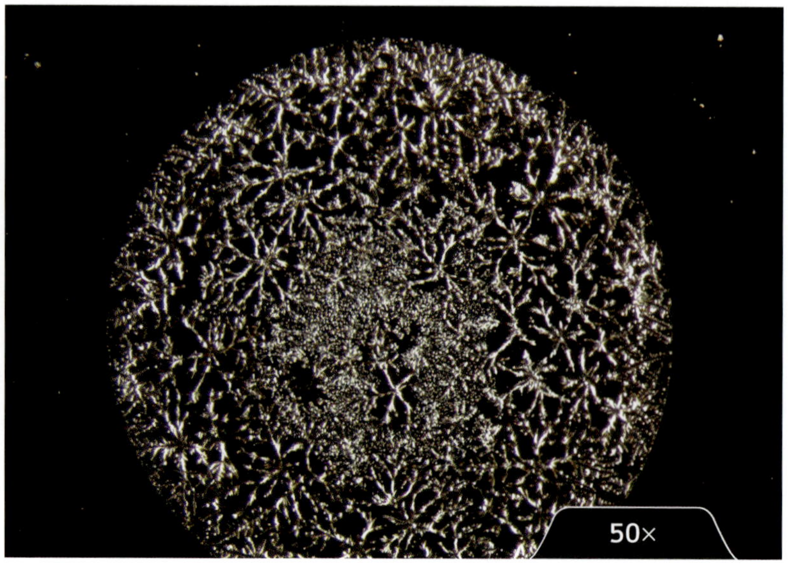

ORGANIC

400×
400×
200×
400×
NONORGANIC
200×
ORGANIC
400×

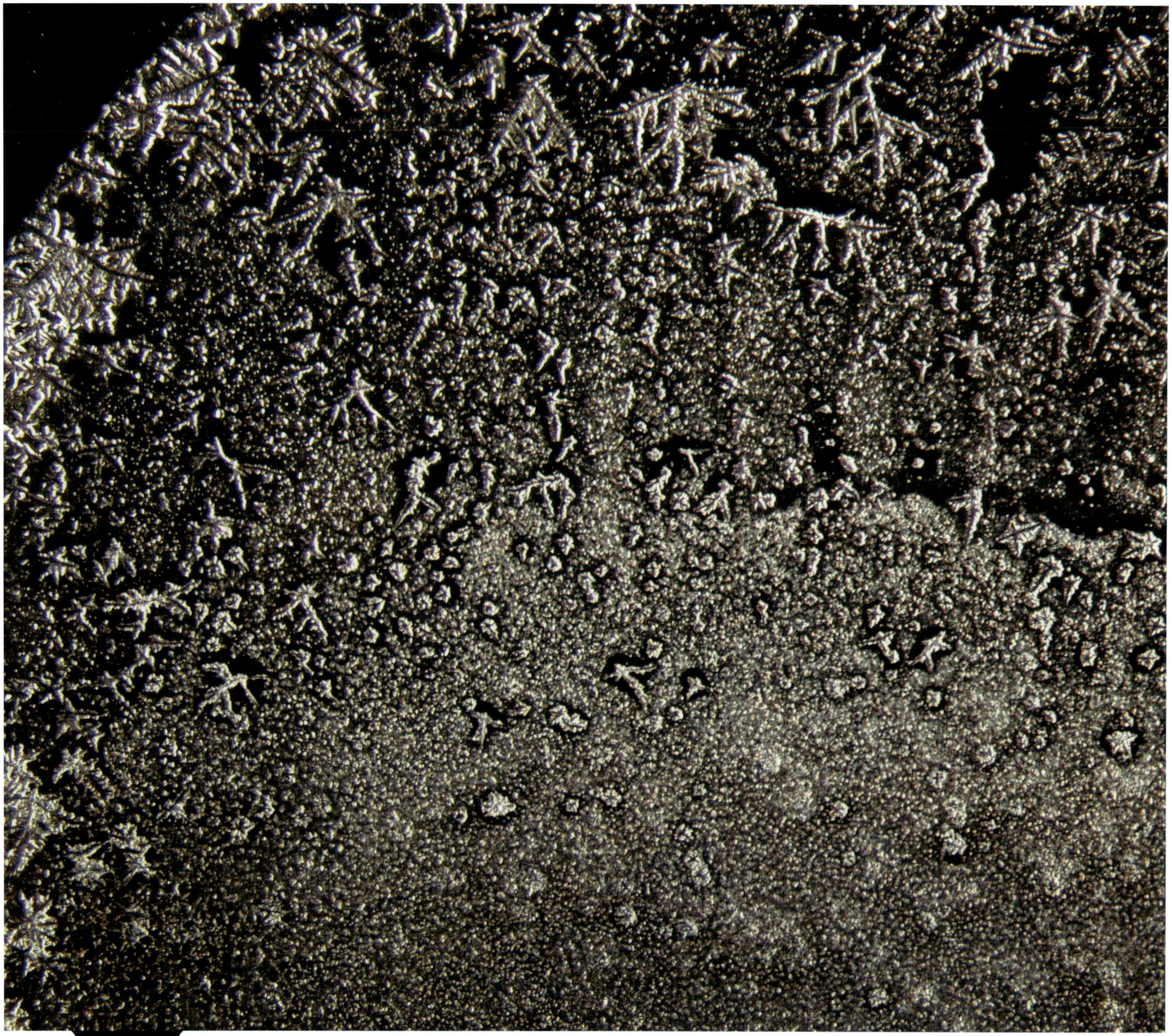
200×

NONORGANIC

200×

ORGANIC

Lettuce

As consumers we just know lettuce as a head of green leaves. We are generally unfamiliar with the actual plant as it flowers and produces seeds. However, the subtle appearance of the organic lettuce is strongly characterized by the blossoms and seeds of the lettuce, which can be seen here in bright variations. Like big luminous torches or sparklers at Christmas time, they reveal to us the essence of natural lettuce.

The nonorganic lettuce on the other hand is degenerated and stripped of its original natural information, to such an extent that it is no longer visible. The image of desert emptiness and sedimentary debris confirms the deadly process that has occurred here.

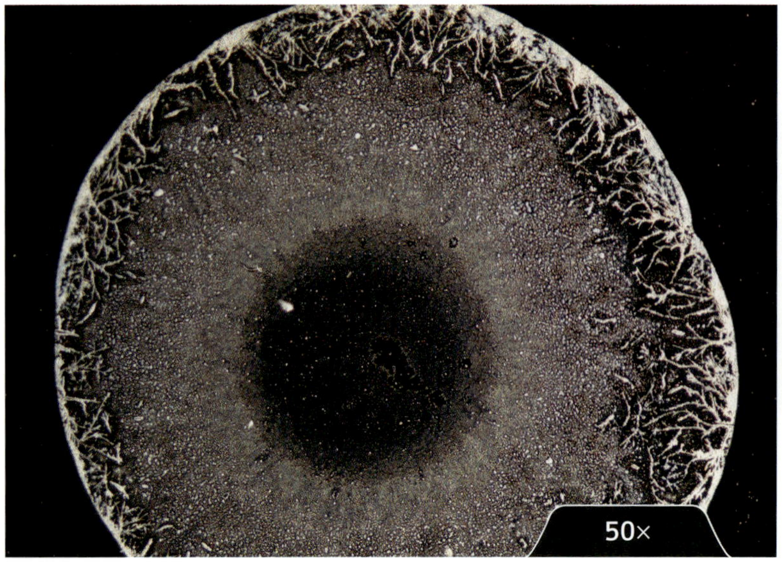

NONORGANIC

ORGANIC

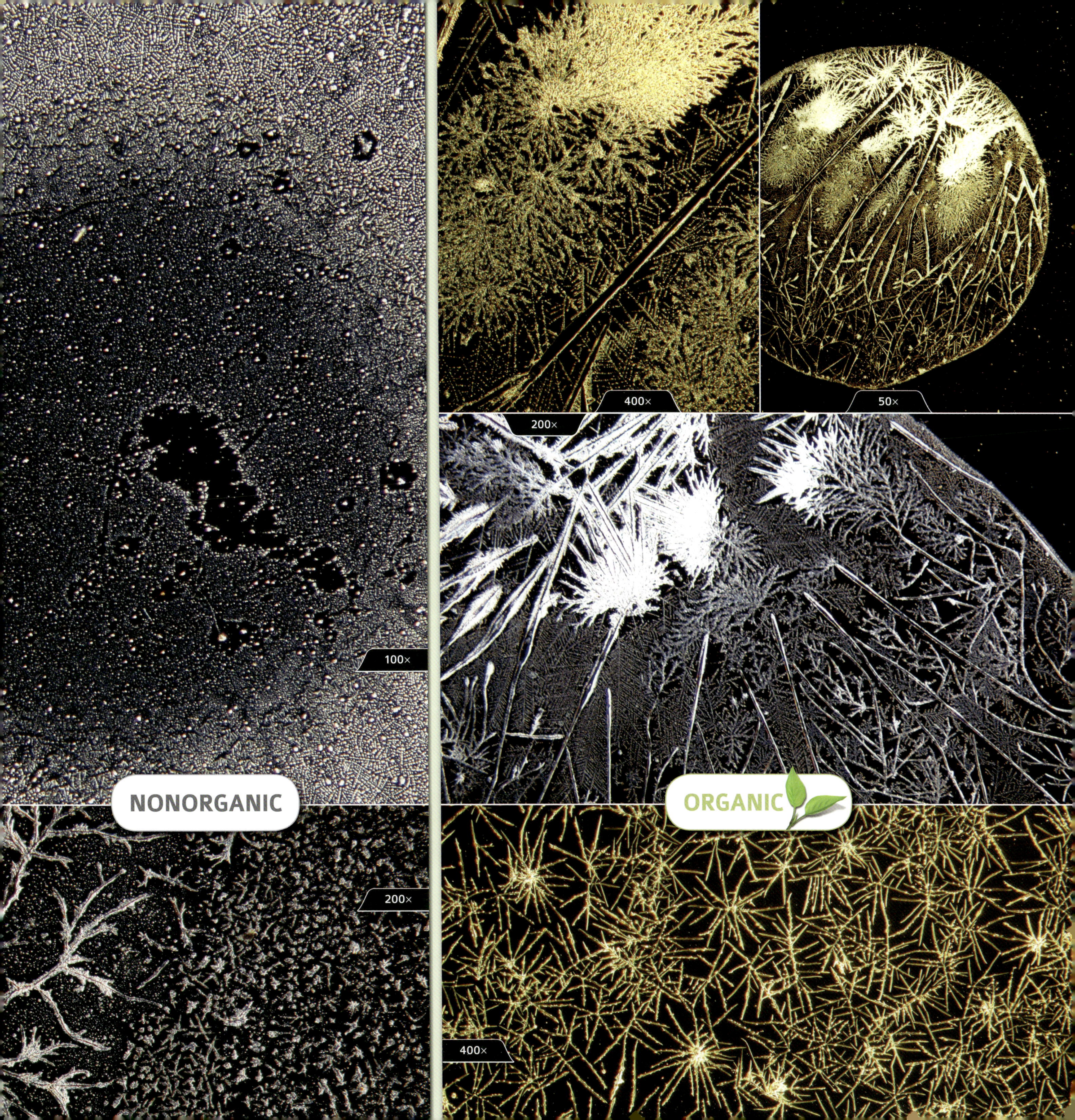
400×
50×
200×
100×
NONORGANIC
ORGANIC
200×
400×

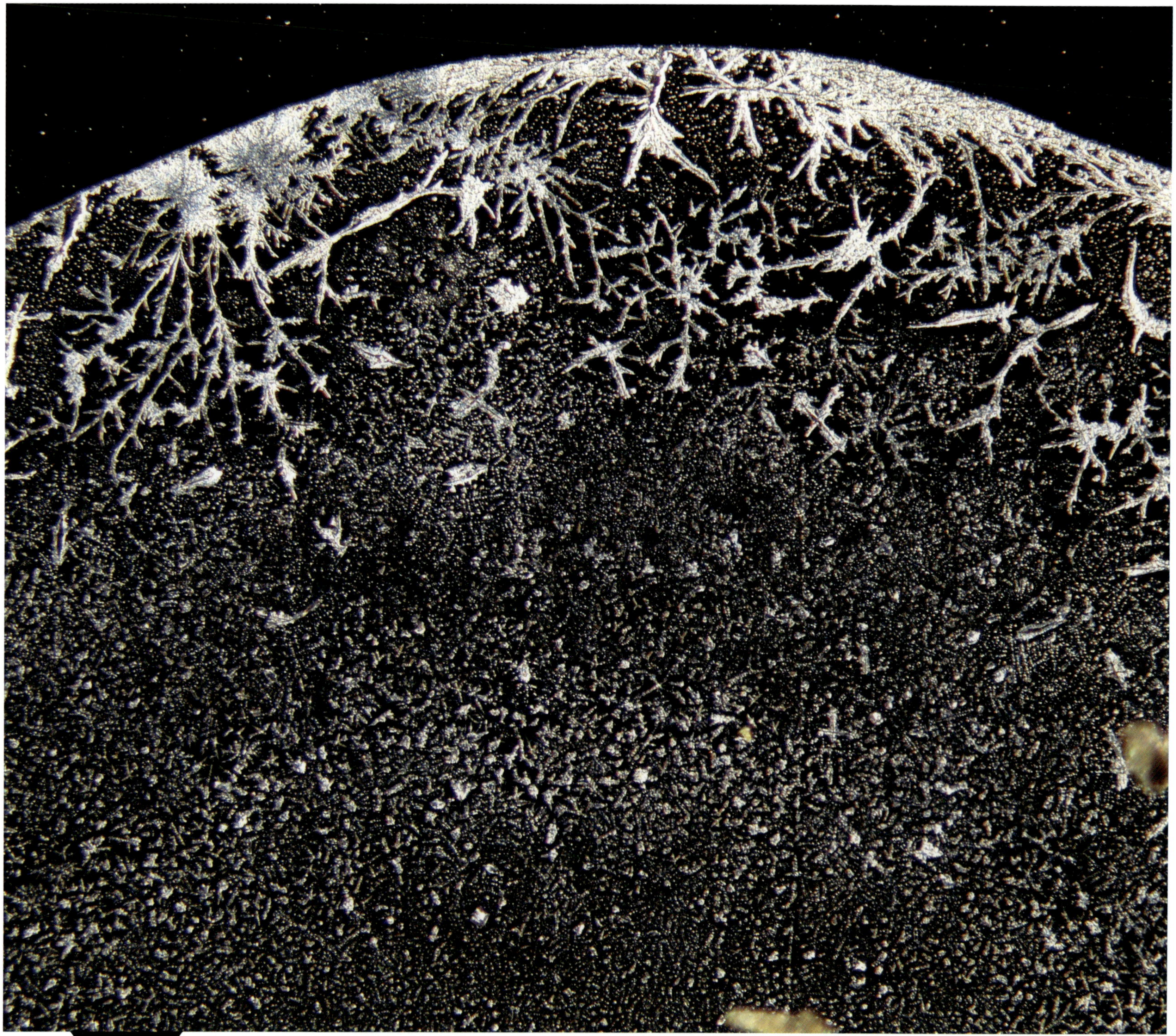
200×

NONORGANIC

200×

ORGANIC

Beans

Surrounded by a wreath of six-petalled blossoms, the organic beans show us an interior made up of a dense and very strong network.

All the more astonishing is the image of the nonorganic beans, almost completely empty of plant-like structures. The wreath is not a wreath anymore and only contains undefined debris and rectangular fragments. Its interior is completely dissolved and slightly sedimented. What sort of information can our body receive from this fine dust?

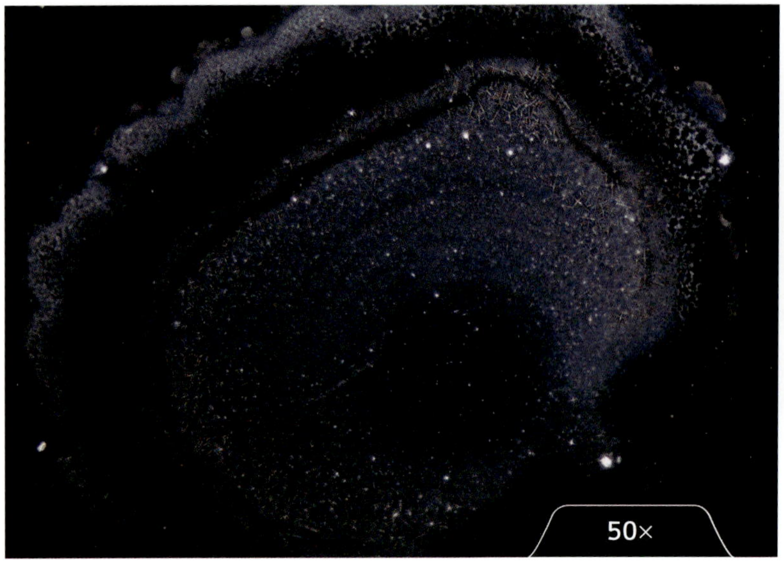

NONORGANIC

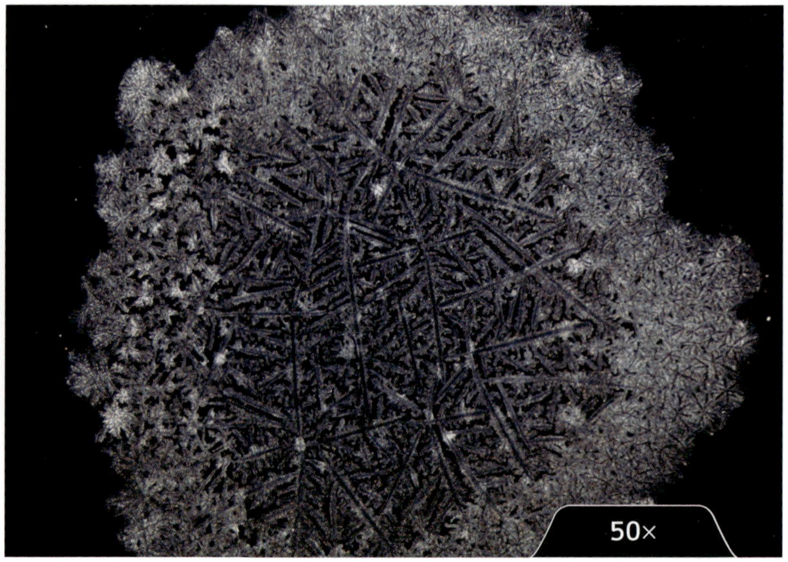

ORGANIC

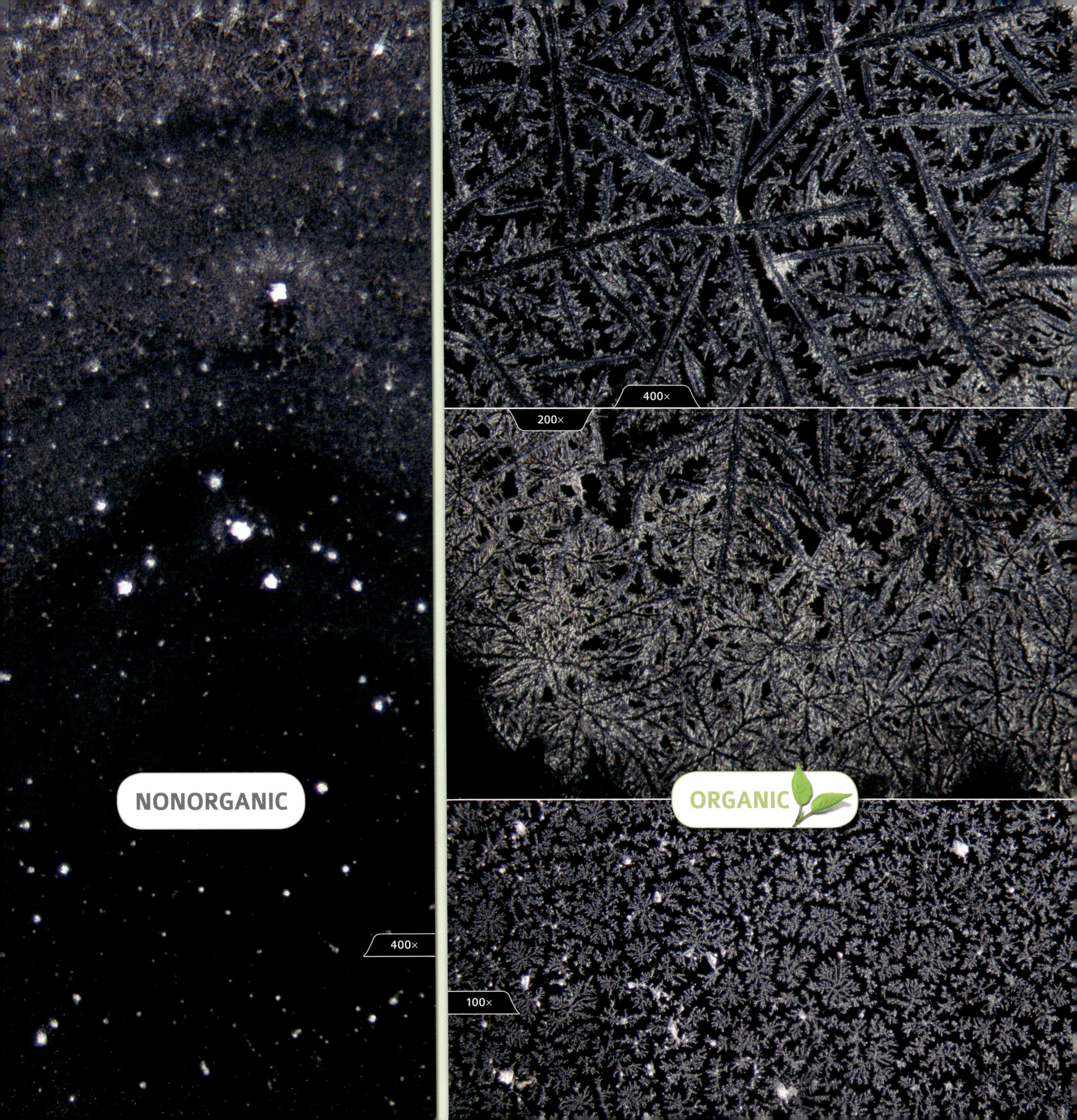

NONORGANIC
400×
400×
200×
ORGANIC
100×

200×

NONORGANIC

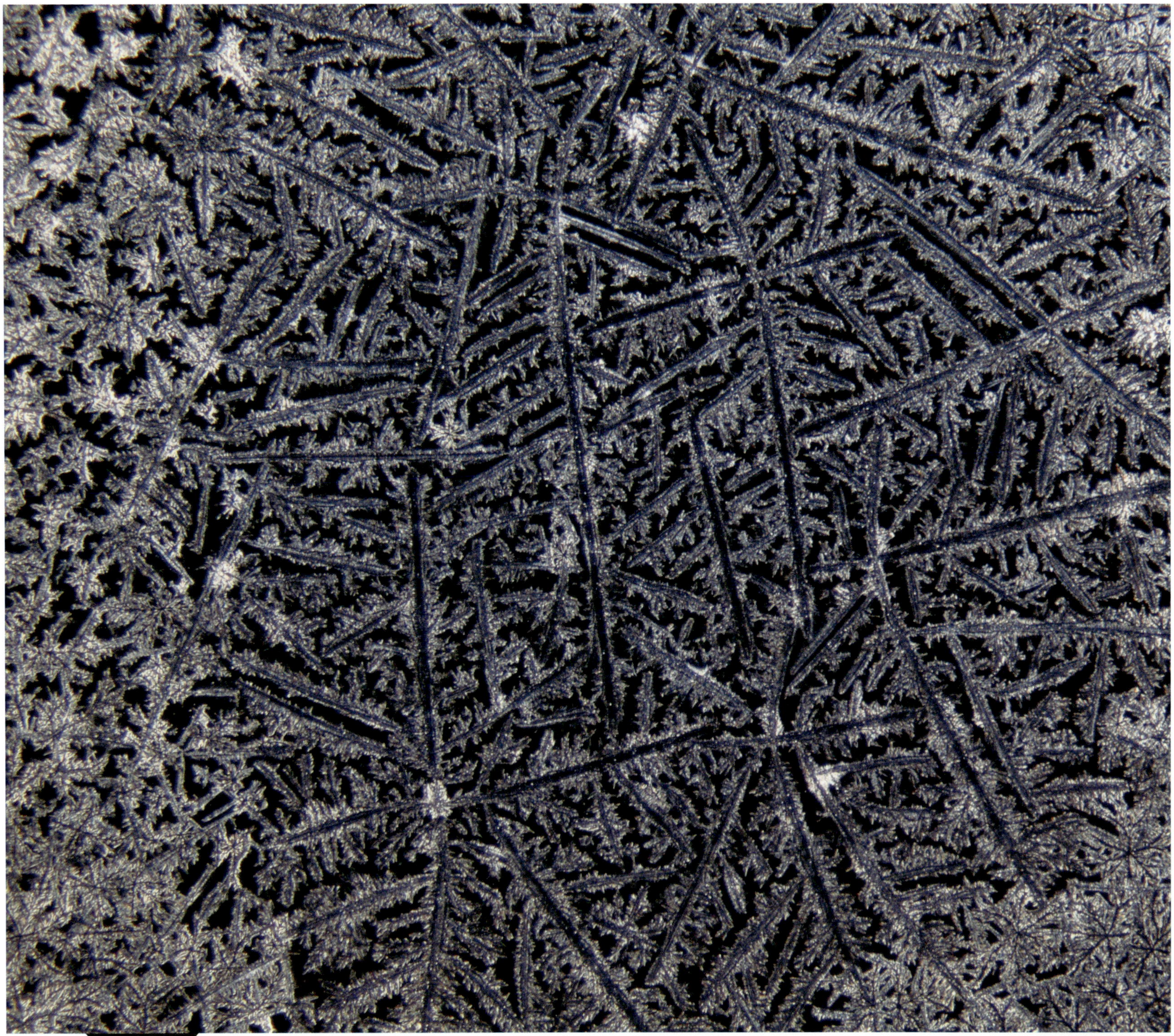

400×

Soybeans

We recognize a very powerful leaf pattern within the organic soybeans, which shows up in the black cut-out shapes, similar to images in some other organic foods. The white shapes are often built from hexagonal elements, very similar to honeycombs. Within the organic soybeans the alchemists' "flower-of-life" manifests itself in another new variation.

In the nonorganic soybeans, no black cut-out forms are visible. The white shapes are very strongly outlined, yet our search for hexagonal honeycombs ends in vain. While the shapes of the organic soybeans have something very spirited, reminding us of glimpses into an imaginary animal world, the nonorganic ones show up as mangled and impoverished.

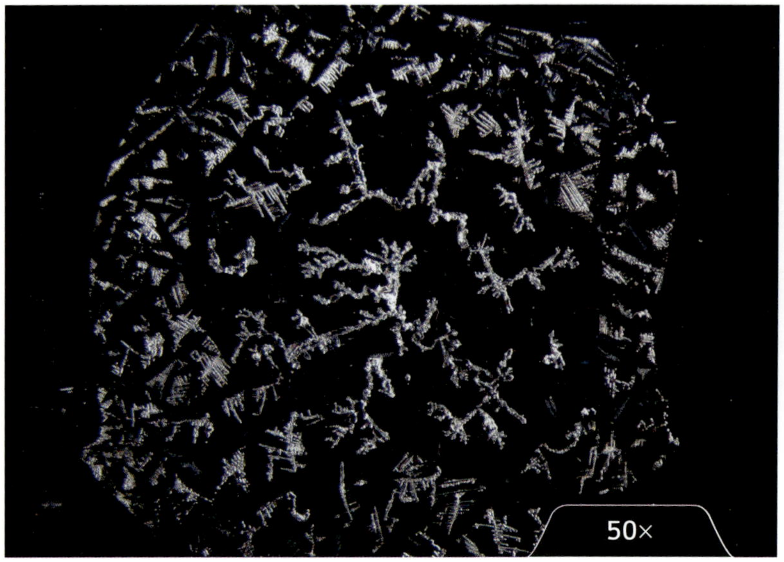

NONORGANIC

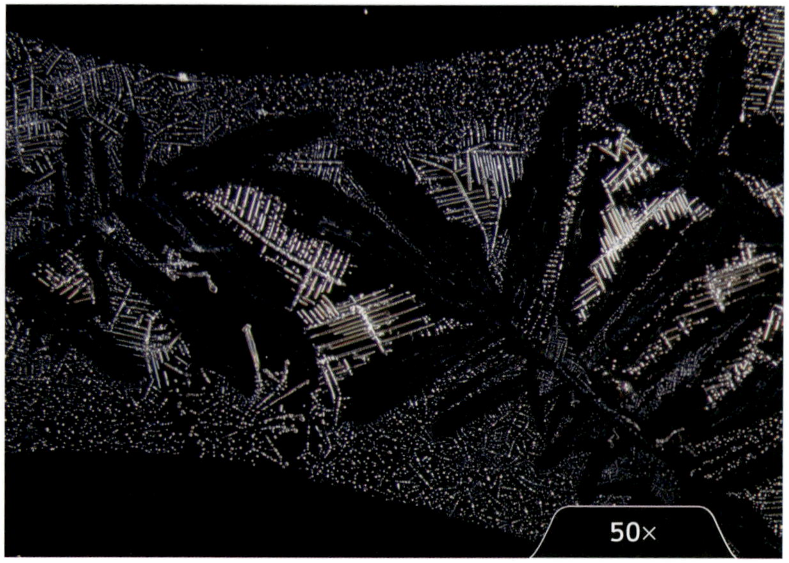

ORGANIC

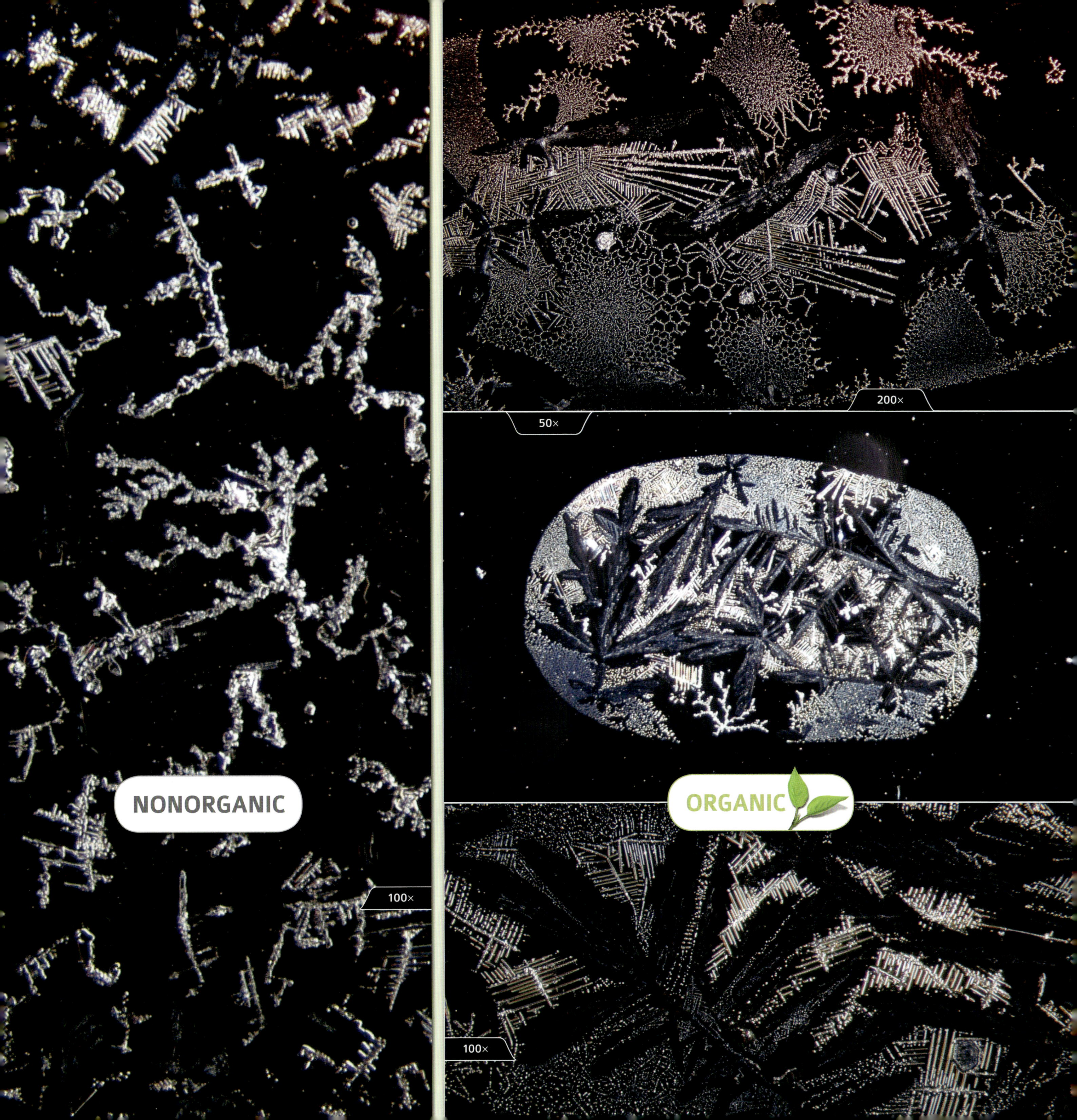

200×
50×
NONORGANIC
ORGANIC
100×
100×

200×

NONORGANIC

50×

ORGANIC

Miso

Miso is an ancient Japanese traditional food, a soy paste fermented over long periods with koji as a starter. The Japanese use miso for their daily miso soup and for dishes with rice, tofu and fresh herbs.

The miso studied here is a barley miso (made with a blend of barley and soy). The organic barley miso is called in German "Nama-BioGersten-Miso" and in Japanese "Nama-Mugi-Miso." For many years Soyana has bought it from a highly traditional company in the southwest of Japan, 1000 kilometers from Tokyo. This miso is fermented in barrels of cedar wood over two summers. It is unpasteurized, so the digestive qualities of this carefully cultivated food can be supplied directly to the customer. In order to safeguard the valuable enzymes and phytonutrients that arise from the fermentation process, Soyana transports the unpasteurized organic barley miso to Switzerland in a refrigerated container, since the temperatures in a non-cooled container can easily rise to 60-80°C (140-175°F), which would thoroughly pasteurize the miso.

The order force (life design principle) of the nonorganic miso is unable to unite the salt with the soy and the barley, making the typical salt crystal appear like a fortress in a battlefield surrounded by aggressive forces, both in the droplet as a whole and in several enlargements.

The organic miso reveals a remarkable harmony, which not only integrates the salt, but also indicates an extremely strong, and sound creative power.

It is interesting to observe how the life-energy of the organic miso generates artistic images, which belong to the archetypal structures of untouched nature and the culture of creative human beings.

In them we can see the fabric of life and shining sun-worlds.

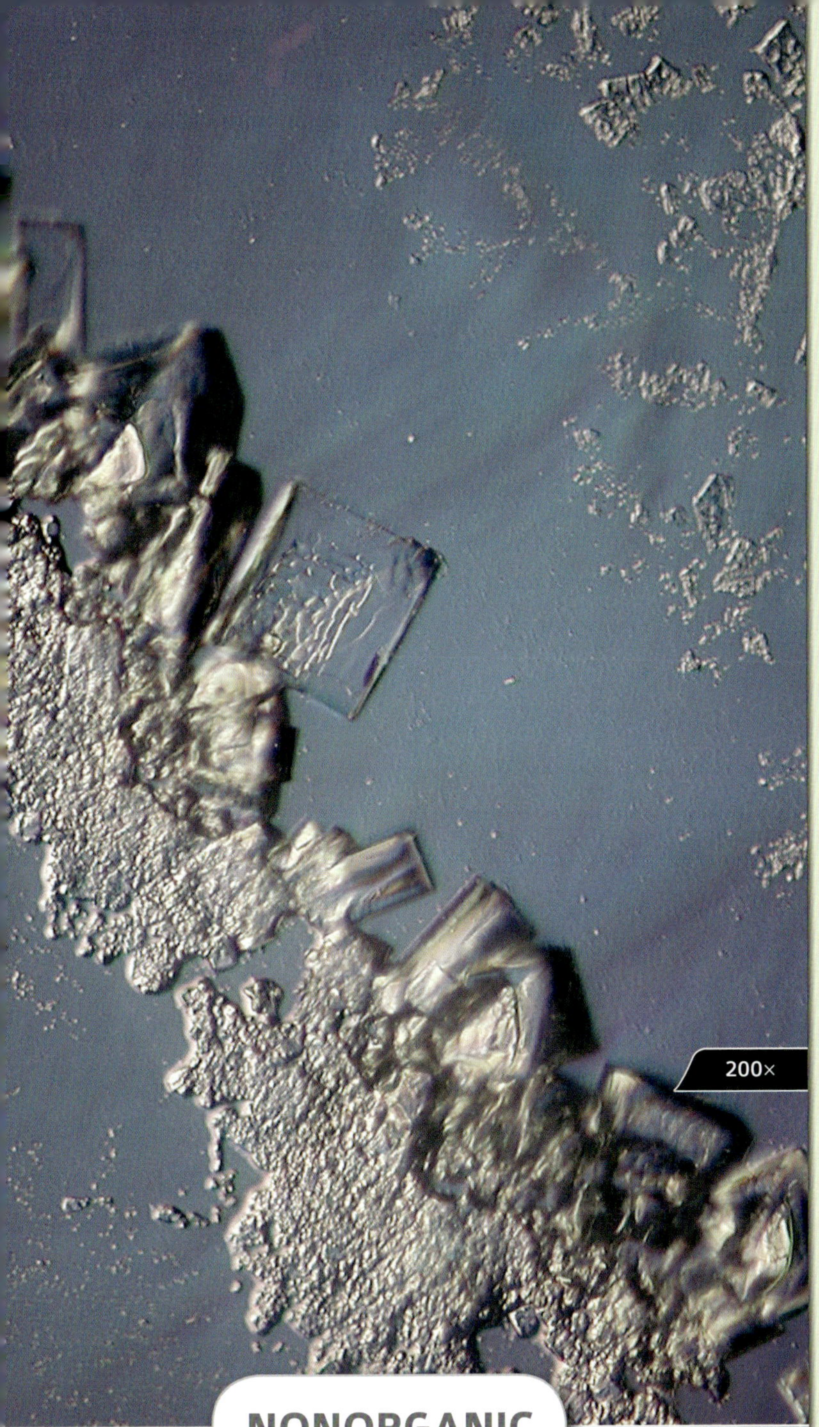

NONORGANIC

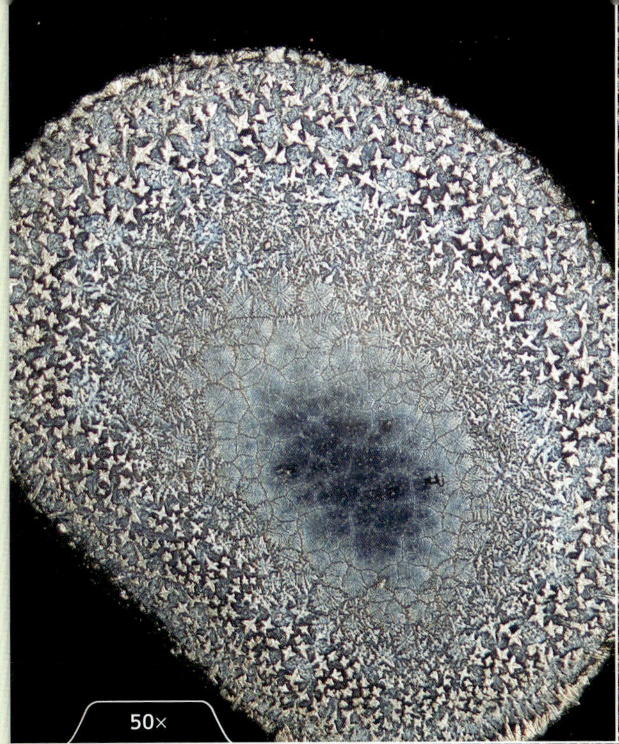

ORGANIC

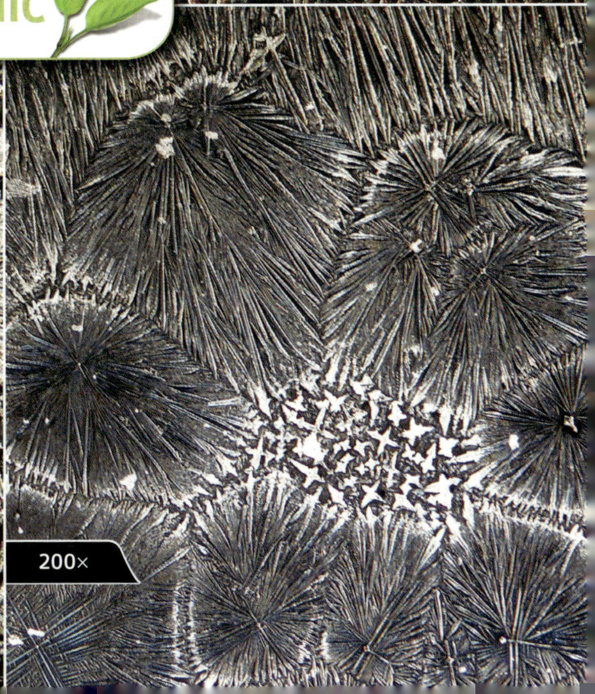

NONORGANIC

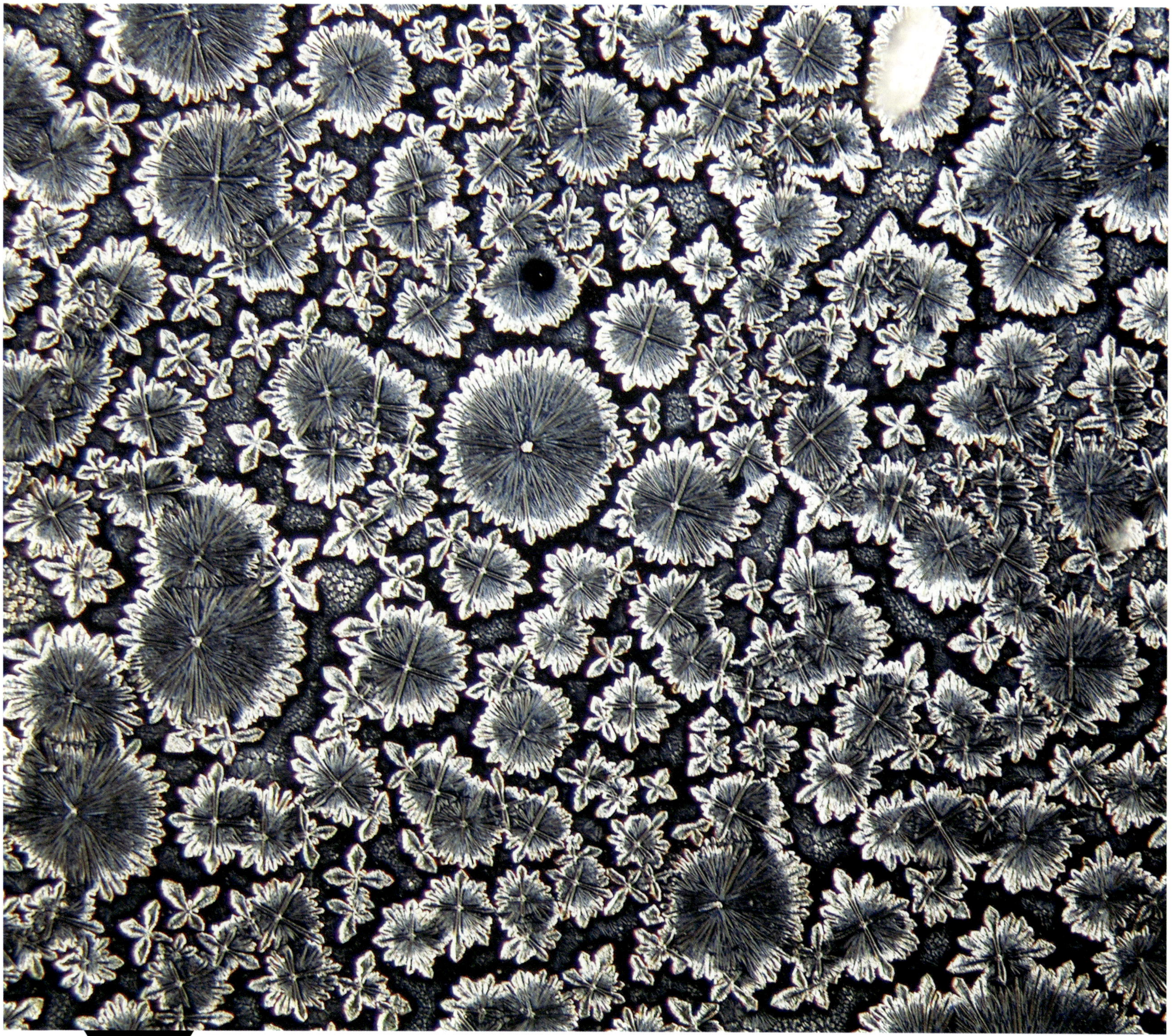

200×

ORGANIC

Tamari

Together with shoyu and miso, tamari is part of the traditional family of the most commonly used fermented soy foods in Japan. Tamari is a soy sauce made purely from soybeans. According to many peers it is the world's finest soy sauce when it is produced organically and fermented over two summers in barrels of cedar wood, sometimes under the open sky, and it can be consumed unpasteurized.

Through this ripening process, the organic tamari achieves a level of tremendous refinement, which can not only be tasted, but now also seen through these images.

Despite a similar ripening process, the structure of the nonorganic tamari remains coarse and unharmonious. It never achieves the consciousness of the organic tamari, which transfers its information beautifully and powerfully, very similarly to the organic miso.

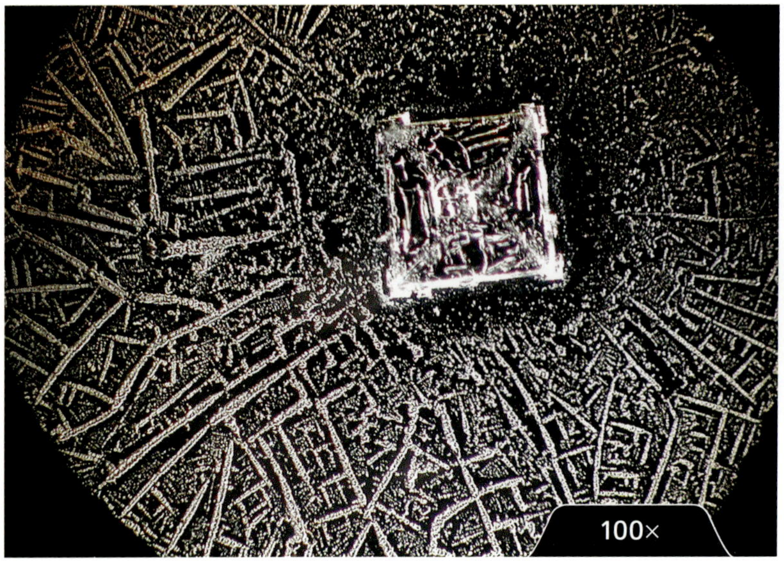

NONORGANIC

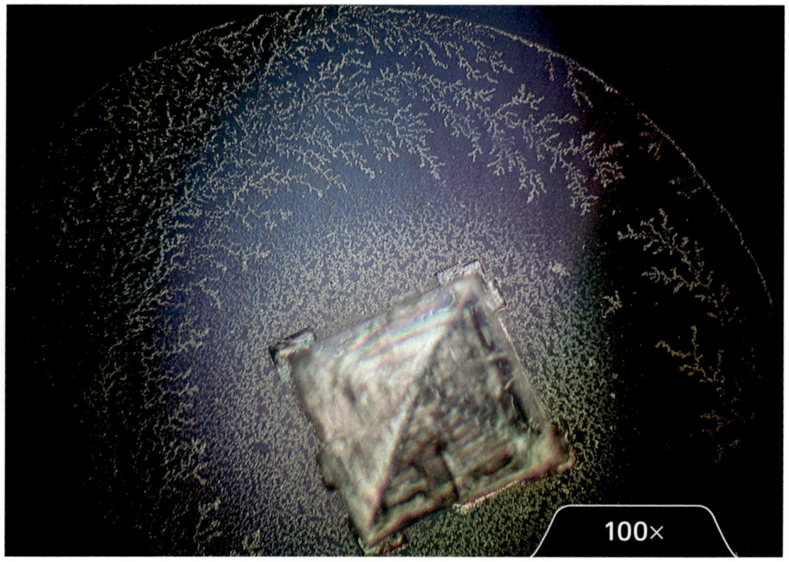

ORGANIC

NONORGANIC

ORGANIC

400×

NONORGANIC

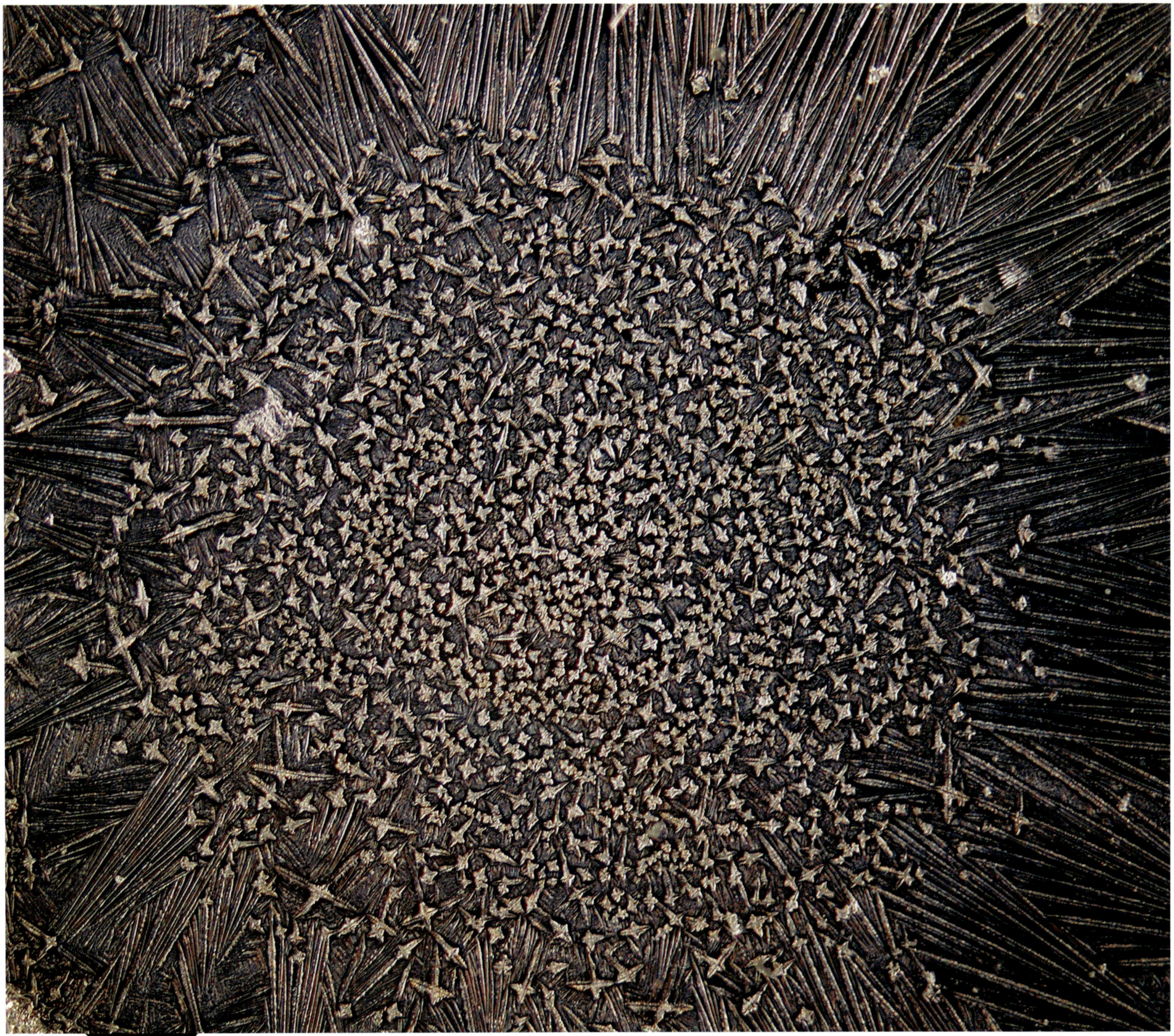

200×

ORGANIC

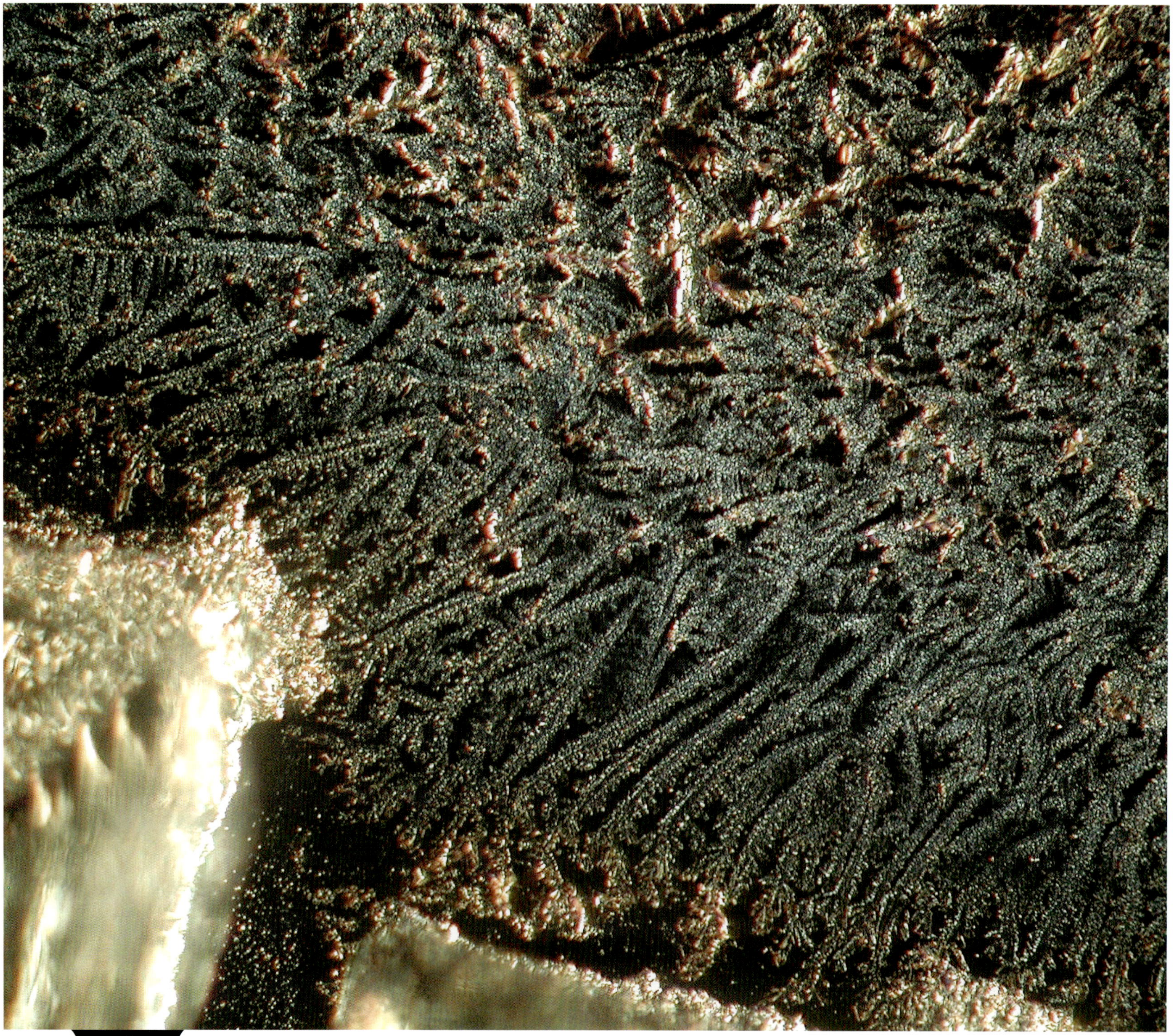

NONORGANIC

ORGANIC

Soyananda Plant-based alternatives to organic fresh cheese

Since at the time of creating this book, there was no nonorganic version of this vegan food available, for once, only the organic versions of this food range are shown. Since, as an observant reader, you will now be amply aware of the obvious defects of nonorganic foods, you can well imagine what the nonorganic versions would look like. Each item of the Soyananda range of foods is organic, fermented and unpasteurized, which implies the highest quality for fermented foods. These were among the reasons which led the German Vegan Society to honor the Soyananda line of Soyana with its innovation award in November 2013. Now, what does the examination of the subtle life-energy in Soyananda foods tell us?

All Soyananda foods have extremely vibrant life design principles, which develop plant-like, powerful and graceful designs. No information gaps are visible anywhere in the drop samples. Extremely lively images can be seen both at the edges and in the centre in expressive and characteristic variations.

Soyananda
Vegan Cream Cheese
100×

Soyananda
Vegan Fresh Cheese,
Herbs & Garlic
100×

Soyananda
Vegan Fresh Cheese,
Sun-dried Tomatoes
200×

ORGANIC

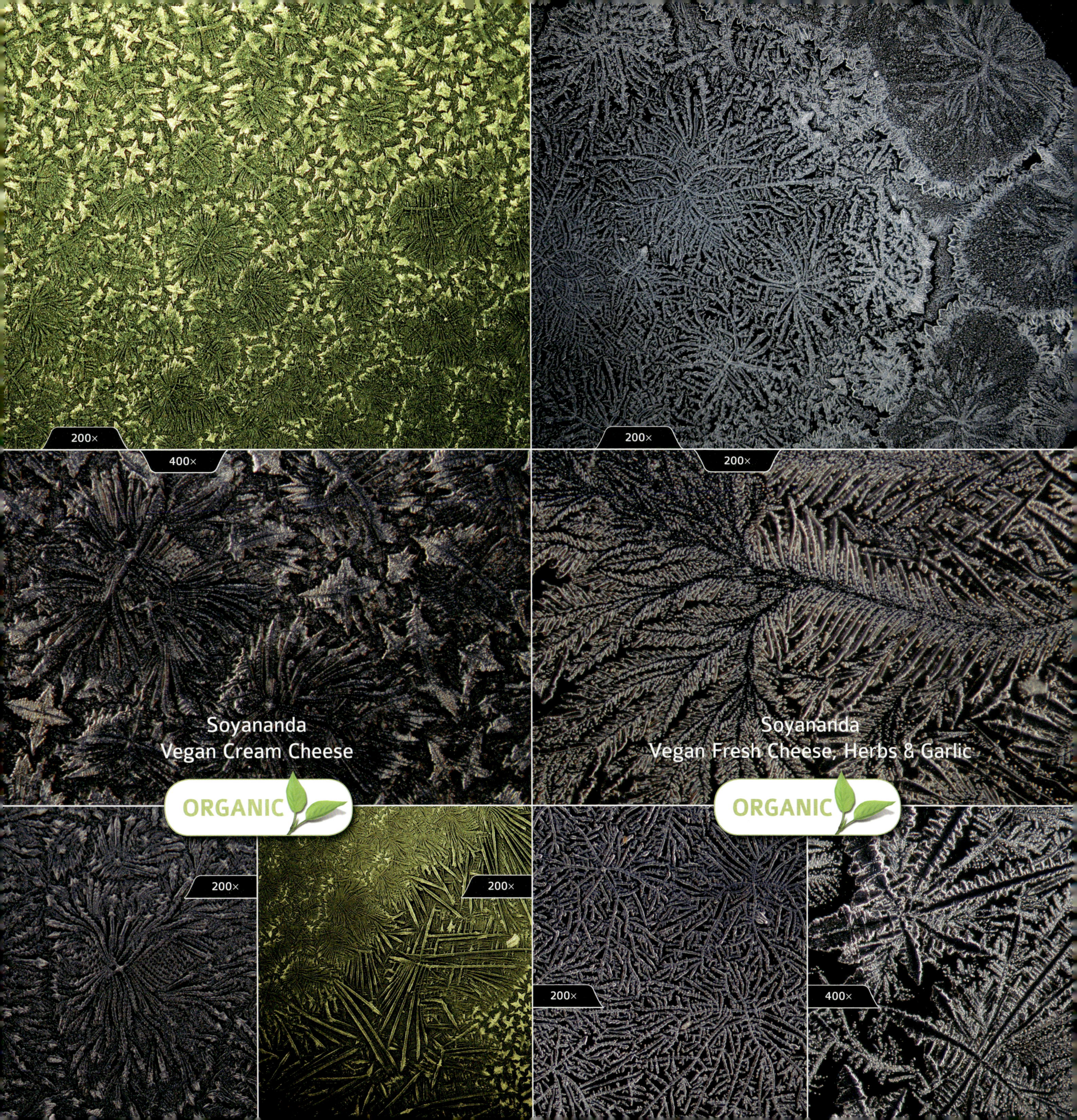
200×
200×
400×
200×
Soyananda
Vegan Cream Cheese
Soyananda
Vegan Fresh Cheese, Herbs & Garlic
ORGANIC
ORGANIC
200×
200×
200×
400×

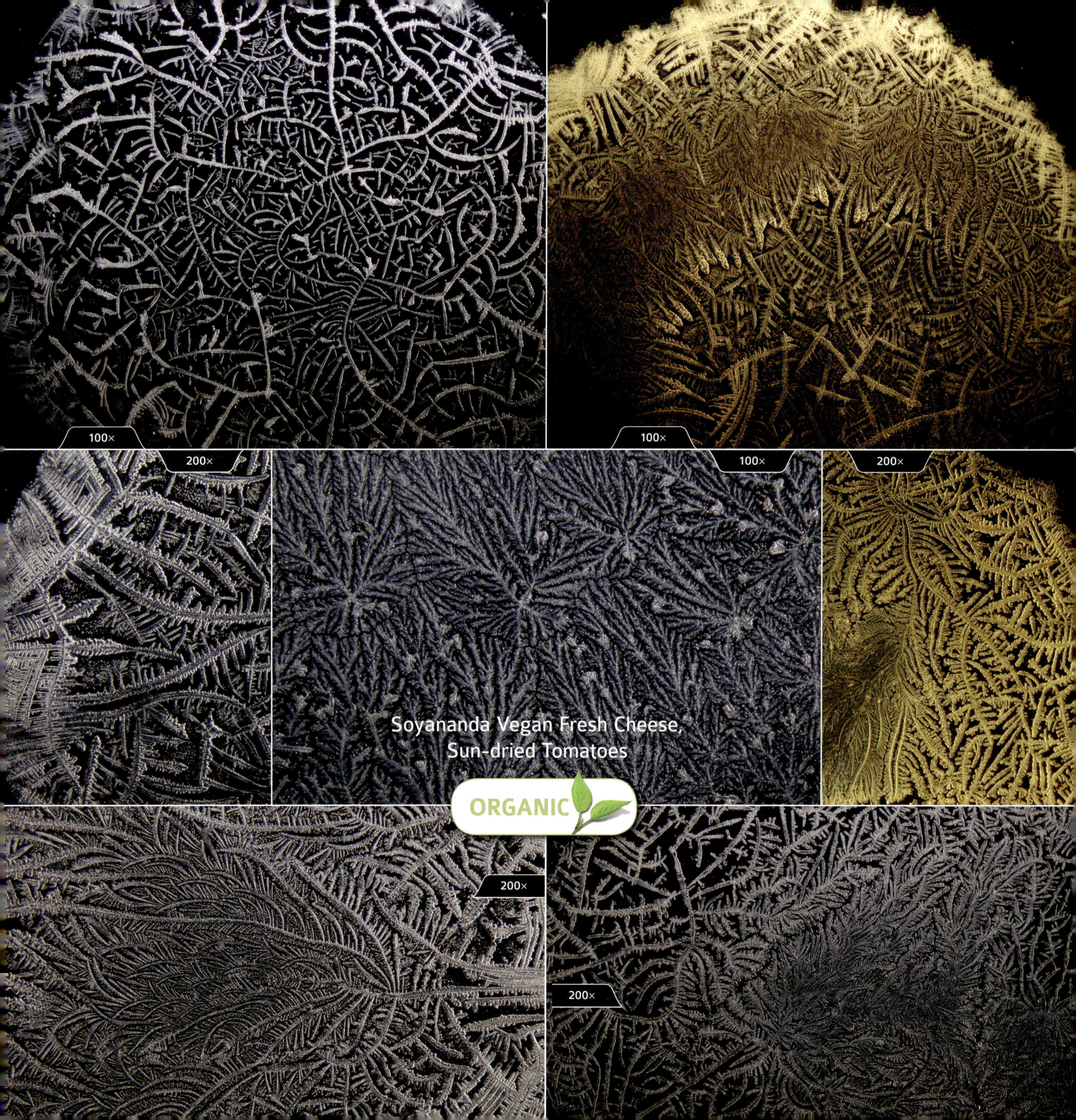

Soyananda Vegan Fresh Cheese,
Sun-dried Tomatoes

Soyananda Vegan Fresh Cheese,
Sun-dried Tomatoes

200×

ORGANIC

Soyananda Vegan Fresh Cheese,
Herbs & Garlic

400×

ORGANIC

400×

Soyananda Vegan Cream Cheese

ORGANIC

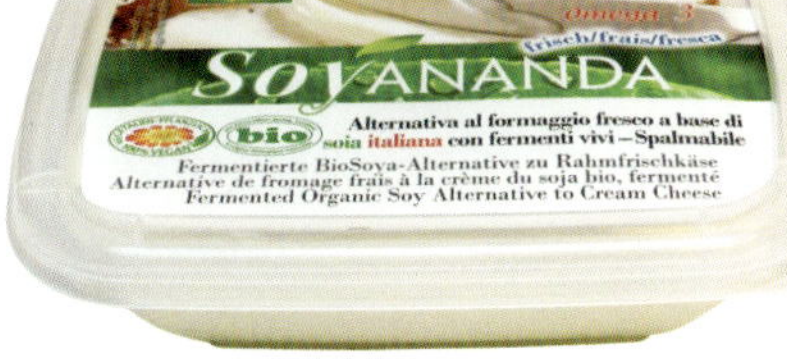

Peas

The entire droplet image of the nonorganic peas reminds us of the nonorganic soybeans, as it also contains a heap of degraded rubble without any beauty and order, and also lacks the black-leaf cut-outs. In other pictures plant-like forms are visible at the edges, although they tend to peter out on the inside and turn into veritable wastelands.

Conversely, the images of the organic peas are filled everywhere with dense plant-like forms, as well as being rife with large six-petalled stars.

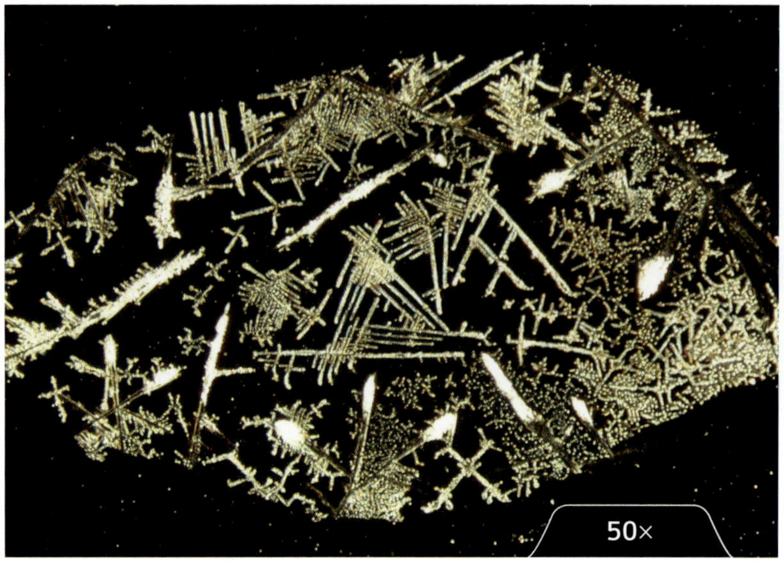

NONORGANIC

ORGANIC

100×
100×
NONORGANIC
200×

400×
200×
200×
ORGANIC
400×

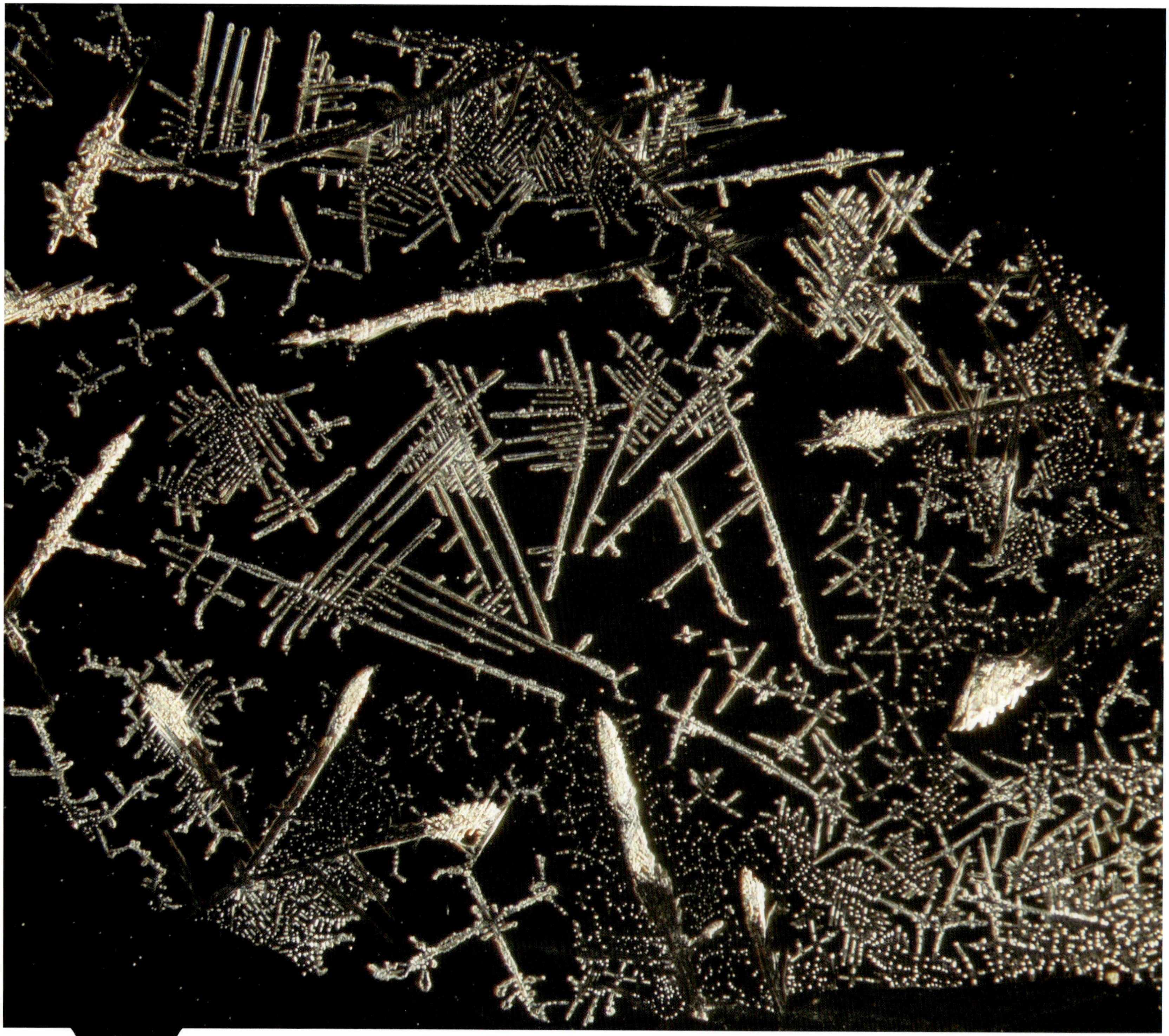

50×

NONORGANIC

250×

ORGANIC

Lentils

The organic lentils gladden us with lushly growing plant-like patterns. Their fullness and style win over anyone who is a friend of natural forms.

An accident must have happened to the nonorganic lentils, since they are almost entirely comprised of information gaps, from which only a few desolate fragments rudely stand out. What a loss for the millions of Indians who prepare their daily dal from such wreckage, compared to the rich information content available within organic lentils!

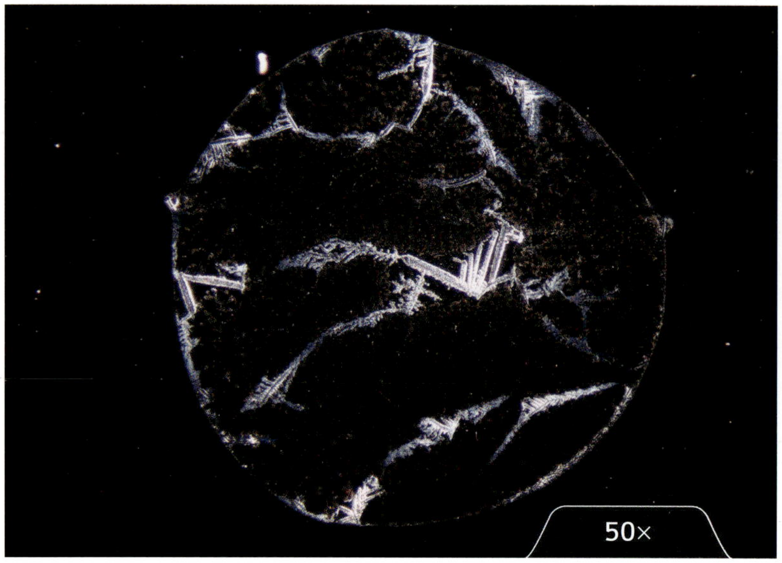

NONORGANIC

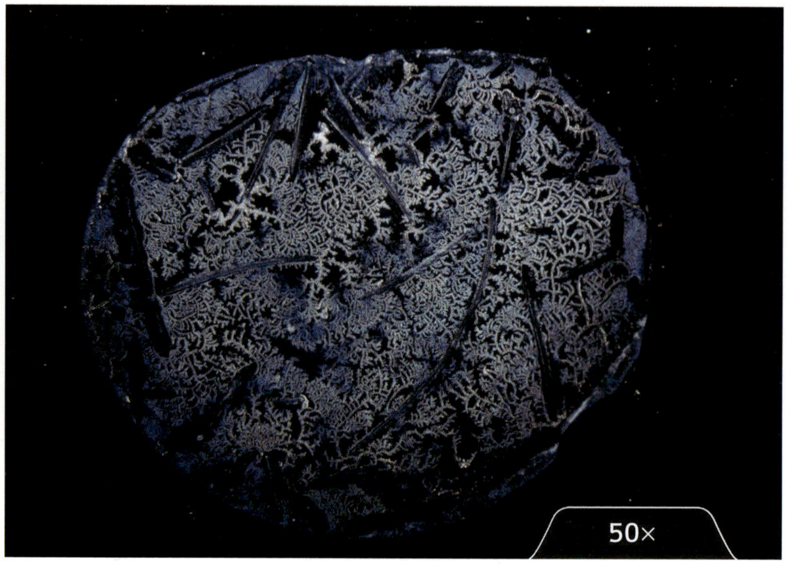

ORGANIC

100×
200×
NONORGANIC
200×
400×
200×
ORGANIC
200×

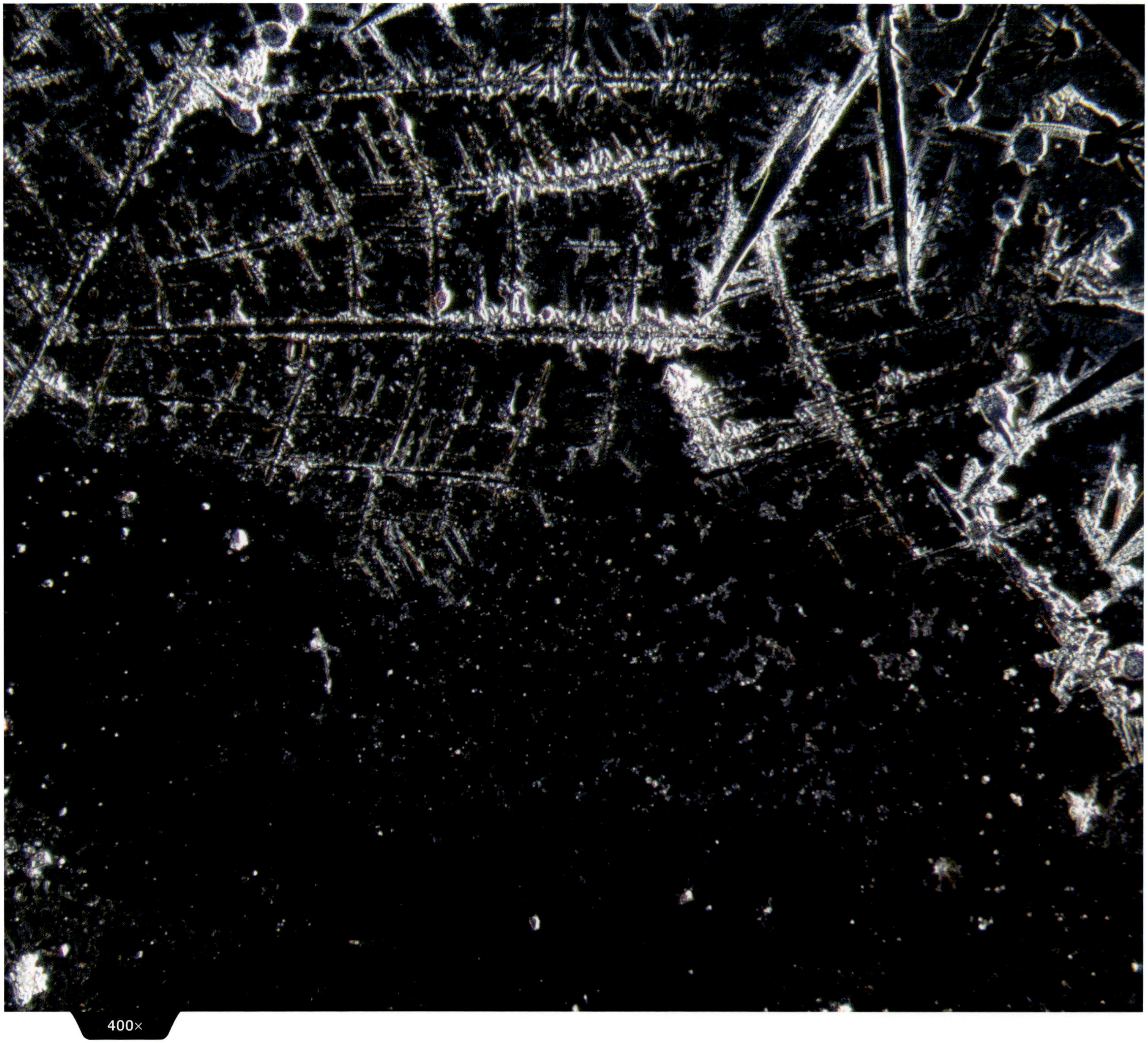

NONORGANIC

400×

ORGANIC

Rice

The organic rice examined here comes from Japan, one of the Asian countries where rice is the daily staple food. The organic rice shows flower-like structures and a tapestry of dense plant-like fabrics.

A catastrophe has befallen the nonorganic rice: it not only lacks the characteristic flower-like shapes, but its field is dominated by information gaps, and whatever is left speaks of degeneration and destruction.

When we think about the billions of people who daily consume such emptiness and messages of disaster, whereas with a little conscious effort they could be taking in messages of living flowers, then we can begin to understand what an opportunity is missed here because of ignorance. We also see the contribution of organic rice to developing and fostering humanity's consciousness and health. May the inspiration of this book lead to improvements.

NONORGANIC

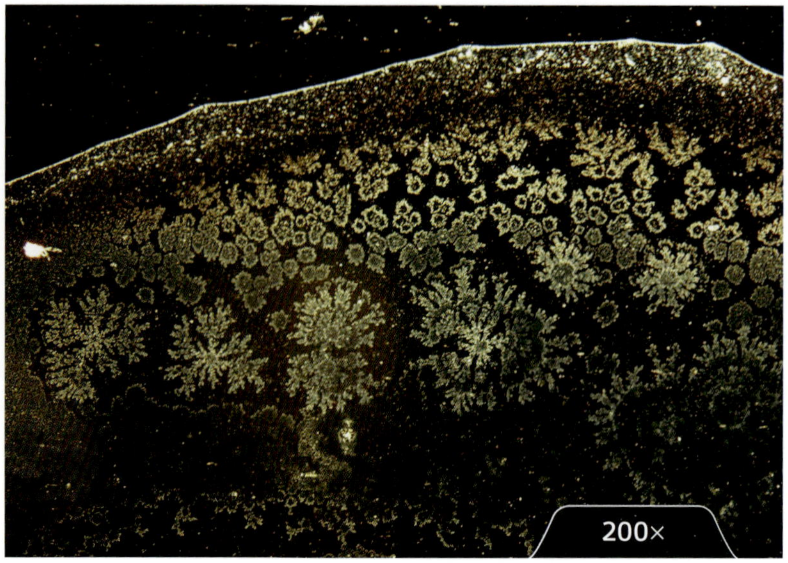

ORGANIC

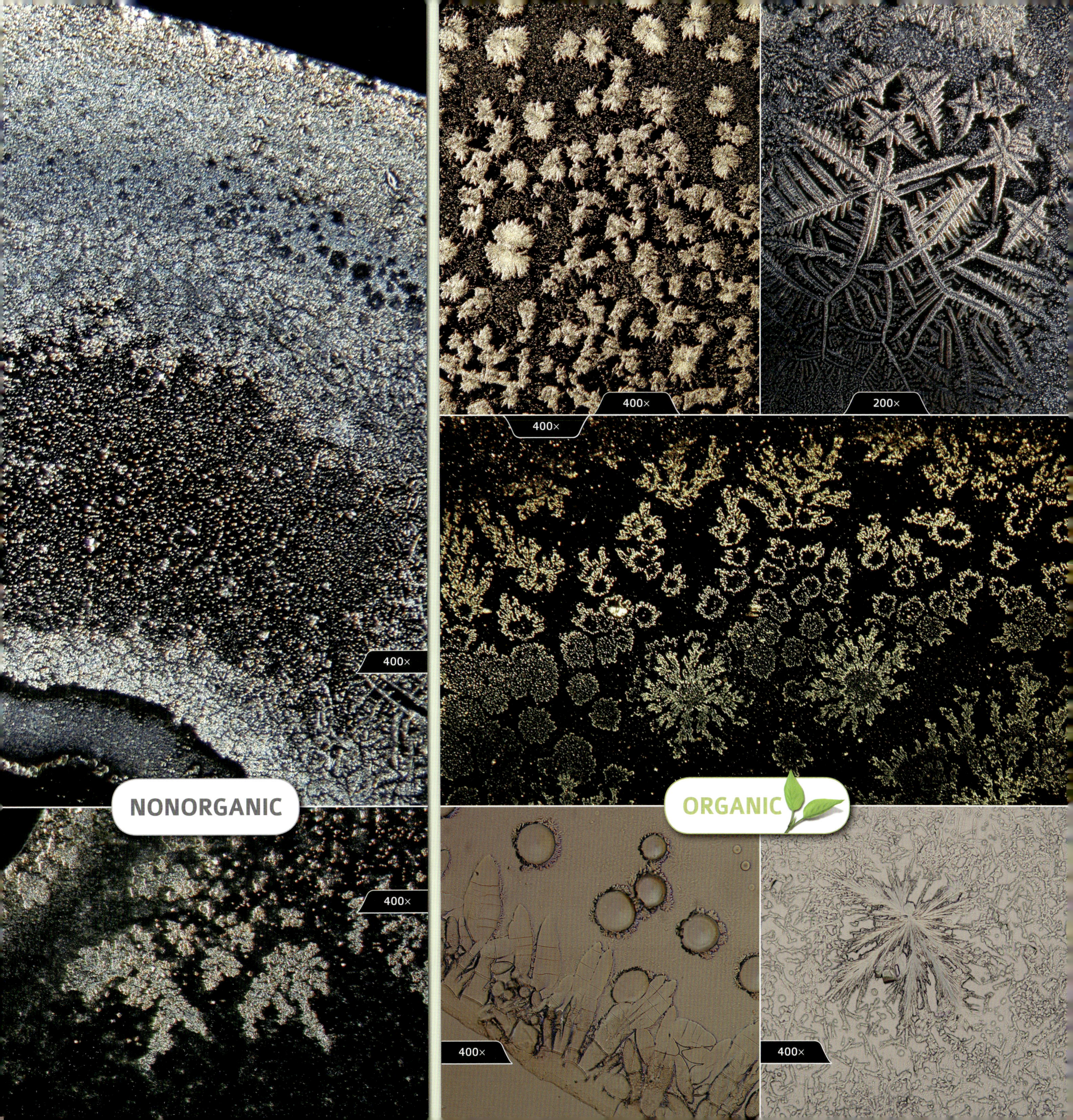
NONORGANIC
ORGANIC
400×
400×
400×
200×
400×
400×
400×

400×

NONORGANIC

400×

ORGANIC

Wheat

The organic wheat reveals its character in plant-like structures and characteristically powerful forms, creating a dense fabric that completely fills up the space.

The nonorganic wheat has almost entirely lost contact with its character and its essential information. Only a few isolated fragments, mostly in the sedimented layers around the edges, give us a vague reminder, but for the most part it is all a yawning emptiness. To lose the order force (life design principle) so profoundly in one of the world's most important cereal grains is nothing less than an agricultural, cultural, scientific and health catastrophe.

NONORGANIC

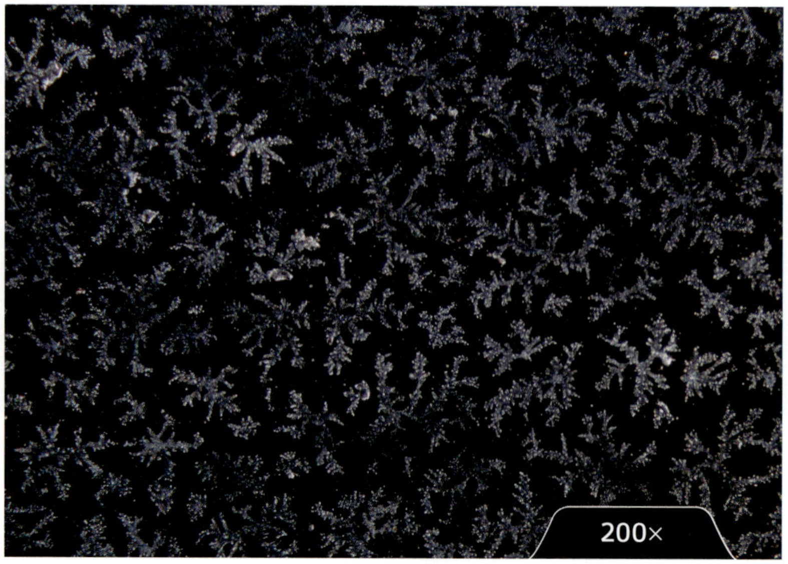

ORGANIC

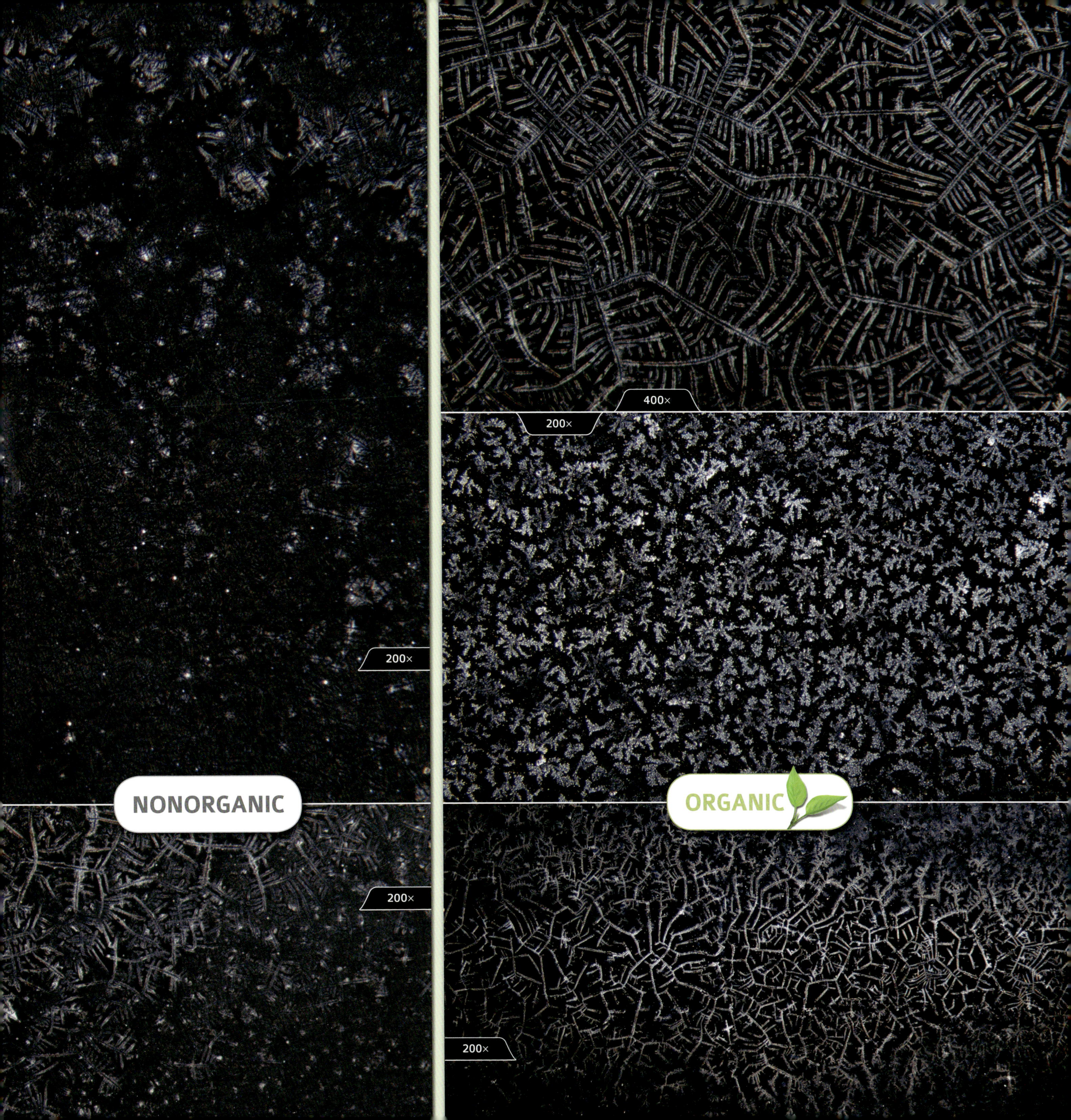
400×
200×
200×
NONORGANIC
ORGANIC
200×
200×

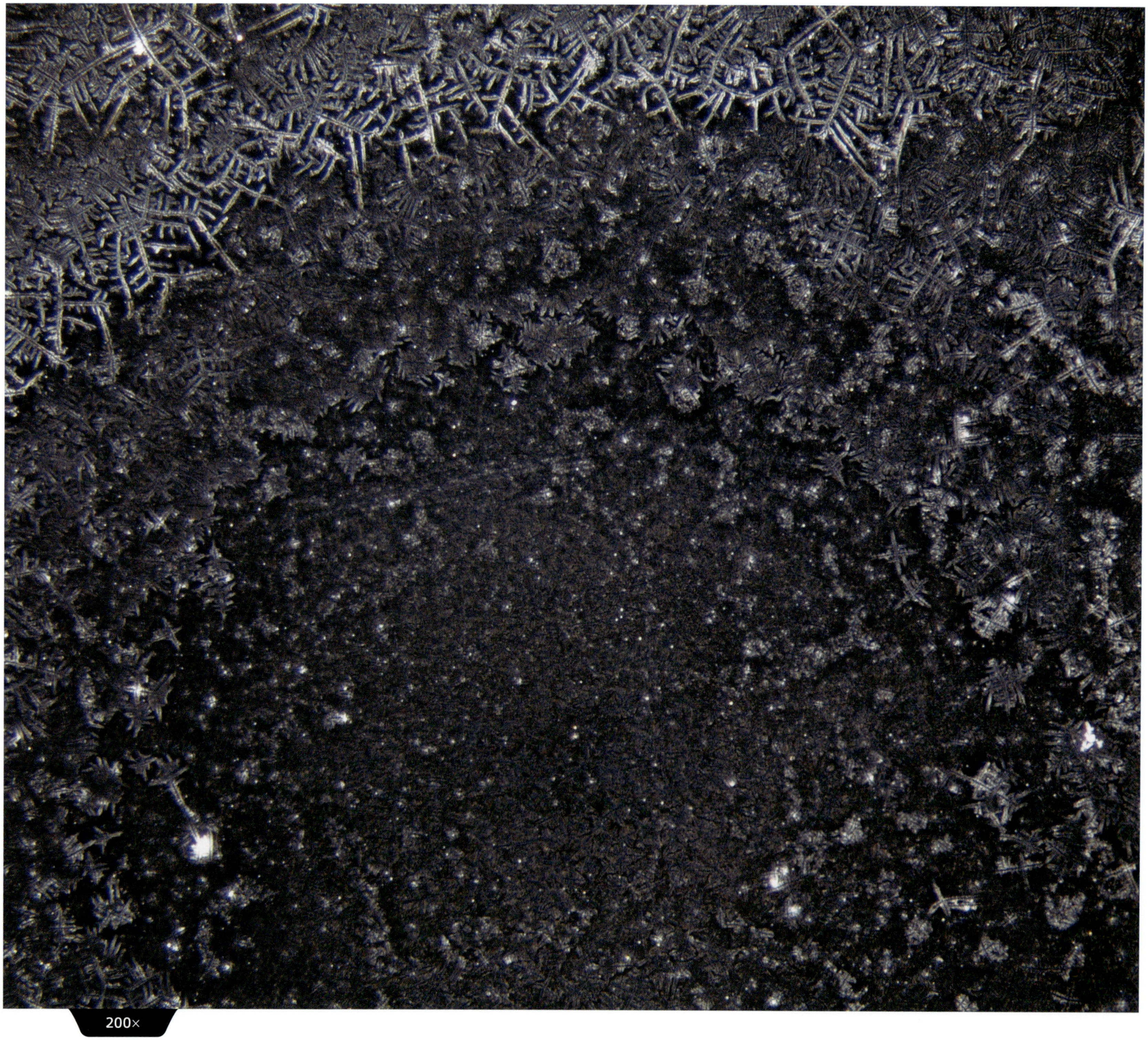

NONORGANIC

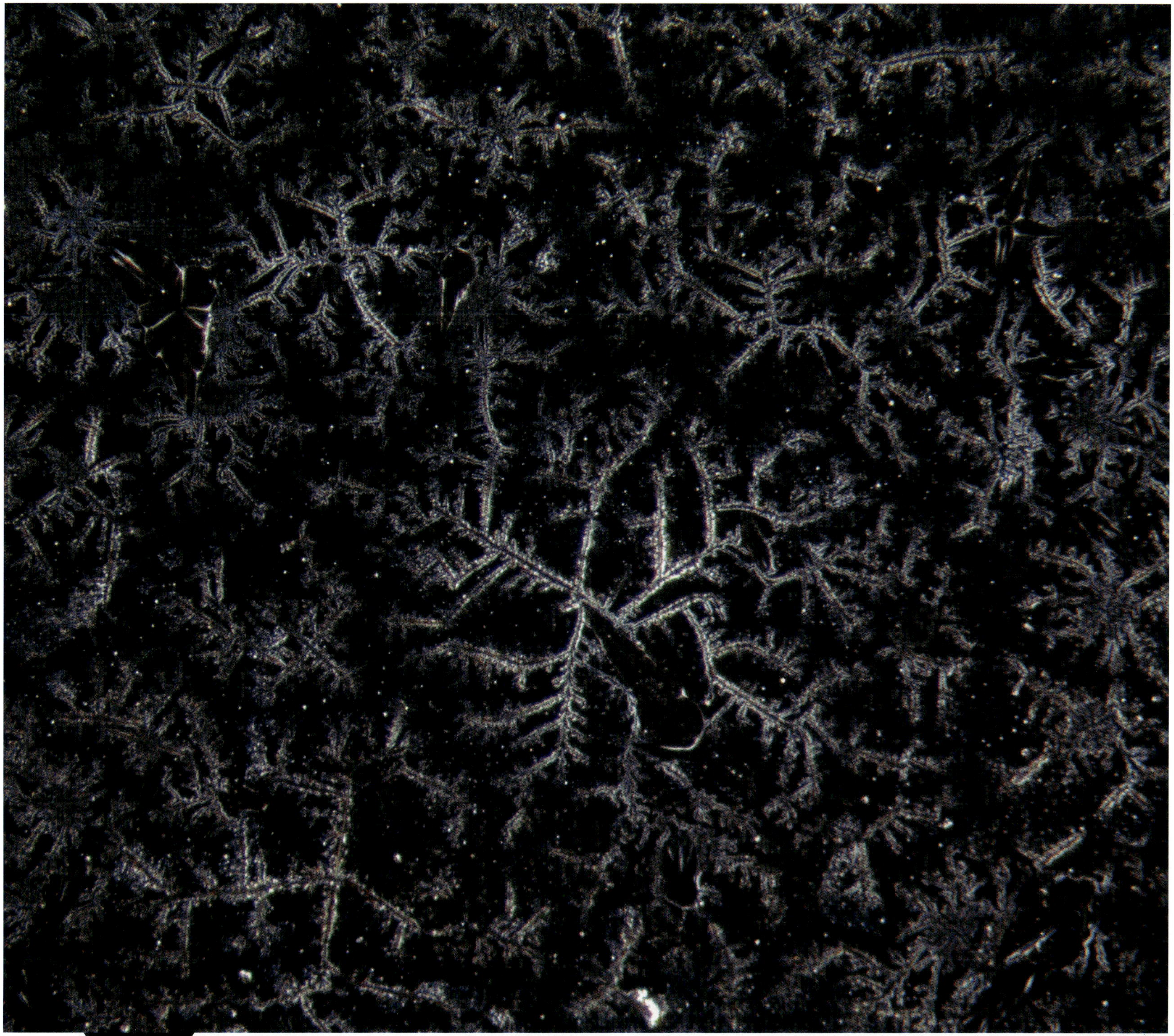

200×

ORGANIC

Pasta (from durum wheat)

The organic pasta fills the space inside the drop with two or more big plant-clusters, often showing a symmetry, whereas the perimeter does not contain any structures.

The nonorganic pasta is empty and sedimented like the nonorganic wheat, permeated by strange and stiff formations. Observing the way the nonorganic pasta presents itself here, it is hard to imagine a food more drained of information.

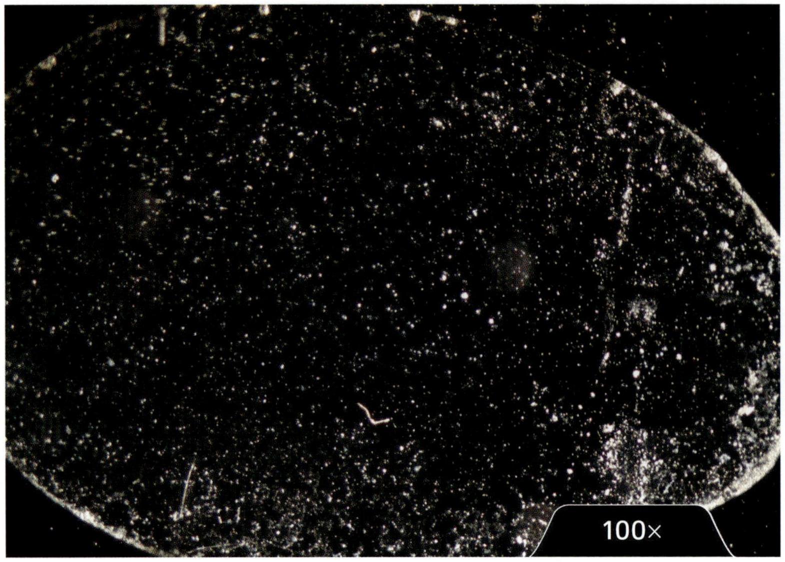

NONORGANIC **ORGANIC**

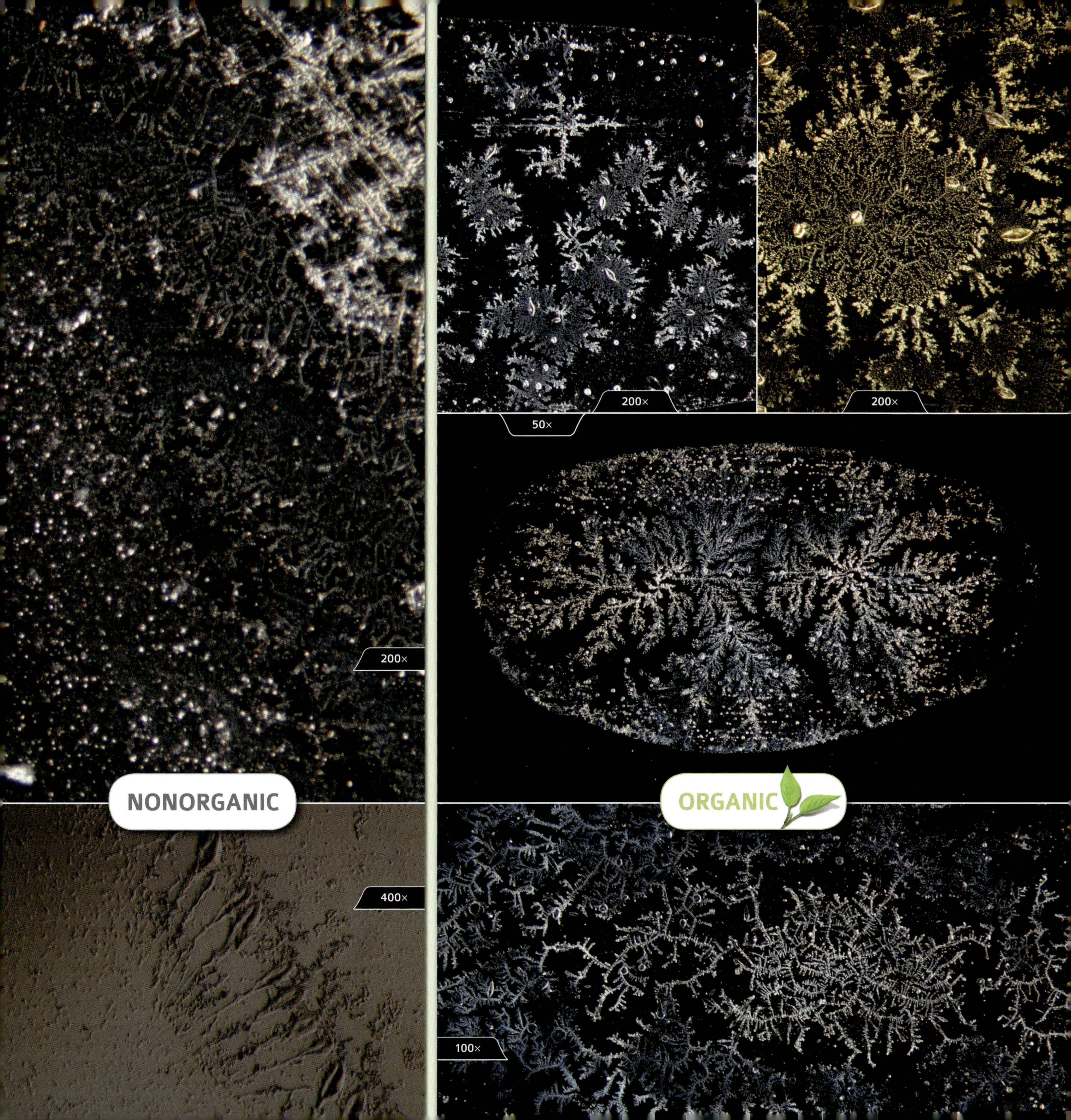

200×
NONORGANIC
400×
200×
200×
50×
ORGANIC
100×

NONORGANIC

ORGANIC

Spelt

The organic spelt bursts with life-energy in various plant-like shapes. It also shows black spear-like cut-outs, which imbue the image with primordial power and form a highly original and characteristic personality.

The nonorganic spelt suffers from different symptoms: sometimes the whole droplet is totally empty of information, like the aftermath of a thunderstorm. And sometimes black cut-outs of rods and beams gather on the edges of an otherwise empty wasteland, a strip of sediment with strong life-signals.

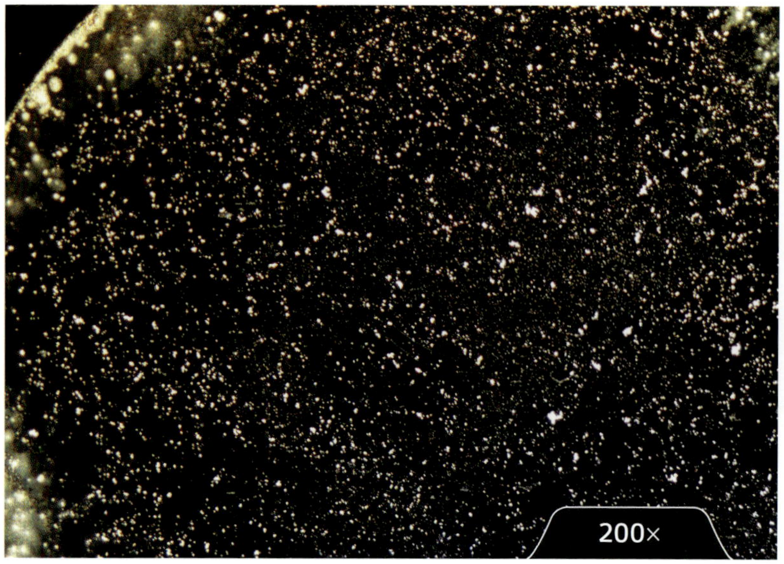

NONORGANIC

ORGANIC

NONORGANIC

ORGANIC

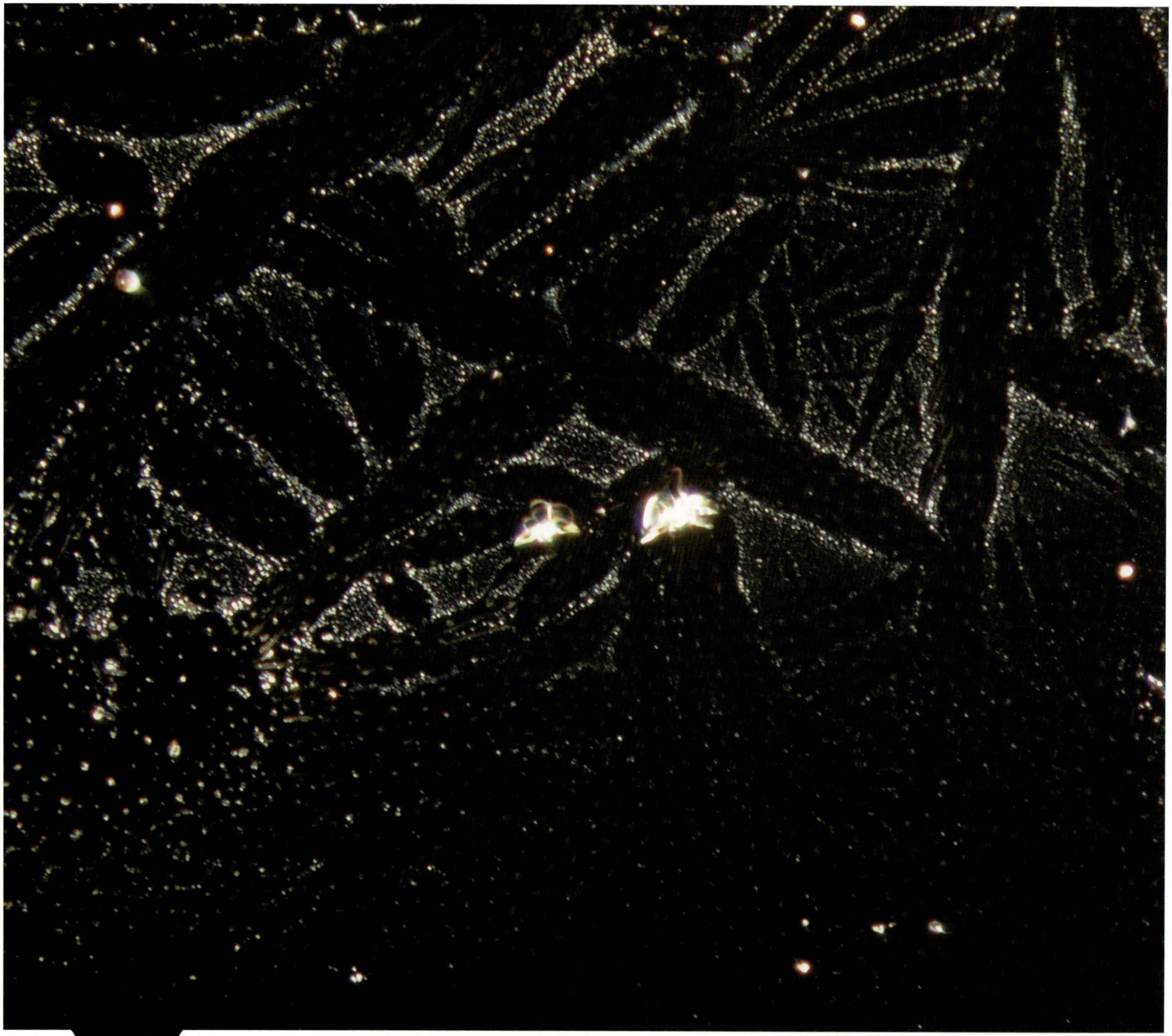

400×

NONORGANIC

400×

ORGANIC

Barley

Characteristically the organic barley is built up from round plant-like shapes, which fill up the entire space. In the larger configurations they reveal solar eruptions in concentric circles; elsewhere they resemble a burning bush, shining powerfully into all directions.

The nonorganic barley has lost this order to the greatest extent possible. Either a bare emptiness yawns or the remaining plant-like elements have forgotten the exact design, managing only wasteful, poorly-defined copies of the original information.

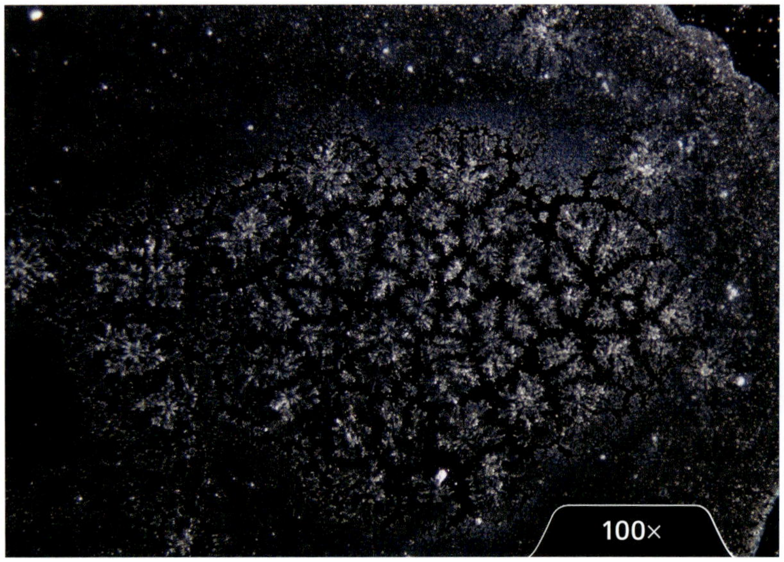

NONORGANIC

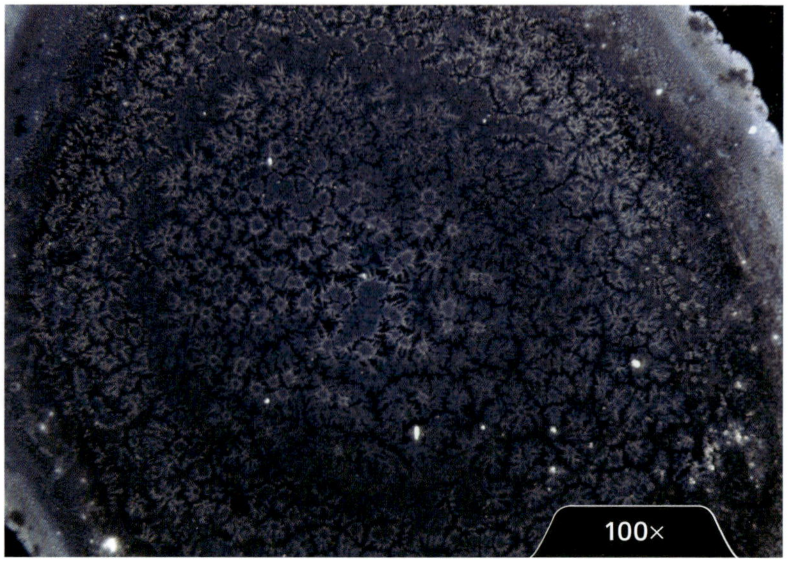

ORGANIC

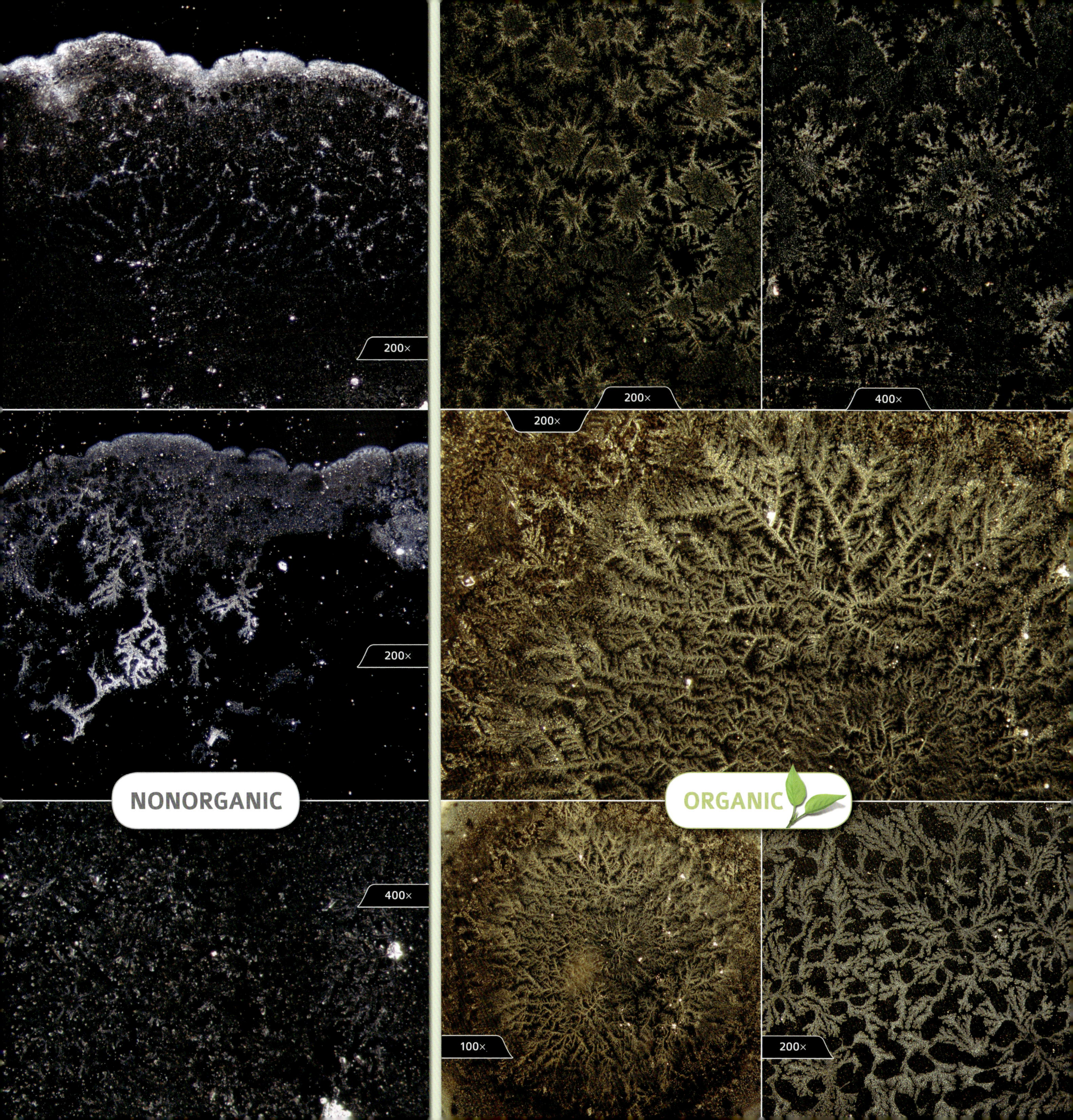

200×
200×
400×
NONORGANIC
400×
200×
200×
400×
200×
ORGANIC
100×
200×

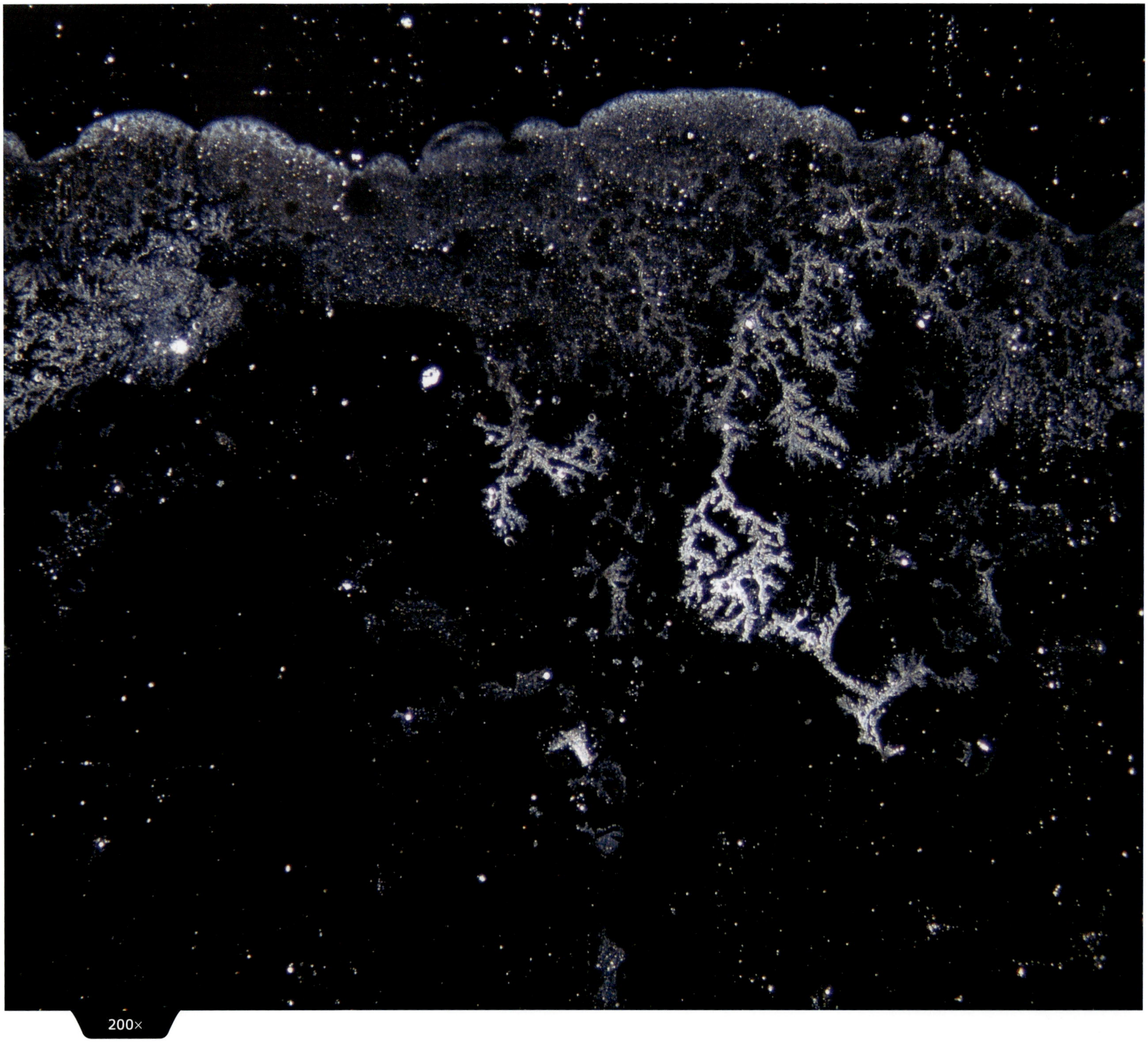

NONORGANIC

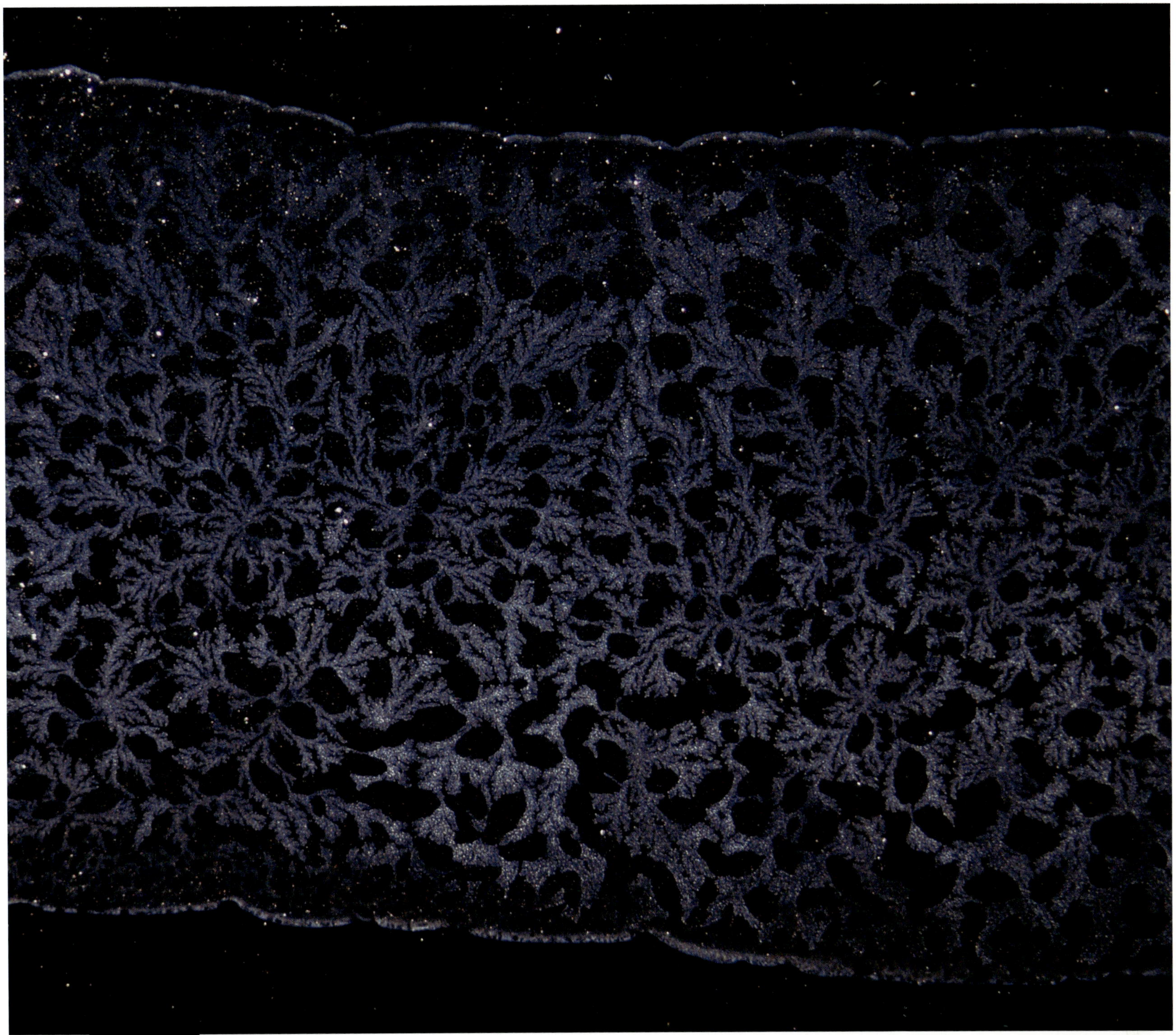

200×

ORGANIC

Oats

The organic oats reveal their character in an incredibly dense and powerful fabric of many shining bushes expanding from a single centre. In the fabric, some single areas are brightly shining, appearing as if lit by a spotlight from within or behind. We have no idea how this phenomenon comes about.

The nonorganic oats are able to maintain part of this order force (life design principle) with beautiful plant-like shapes in isolated areas, sometimes including black cut-outs of plant-like elements. But unfortunately, in most of the drops the order forces of the nonorganic oats push the plant-elements brusquely to the sides, leaving only a big emptiness or a few scattered plant-like remains. The lively fabric which defines the true oats is nowhere to be found.

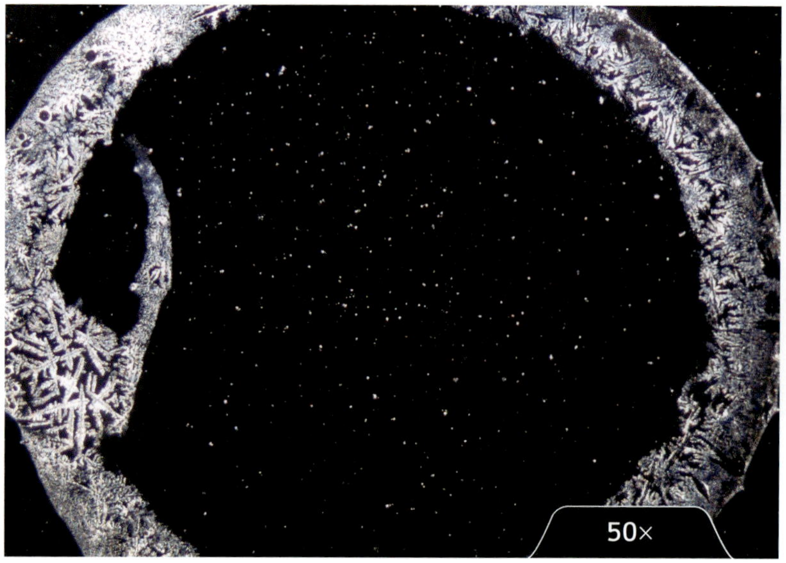

NONORGANIC

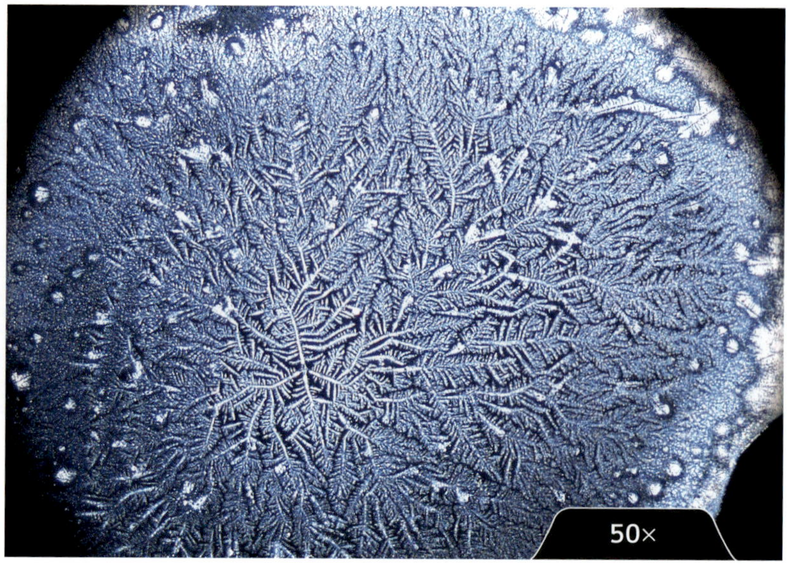

ORGANIC

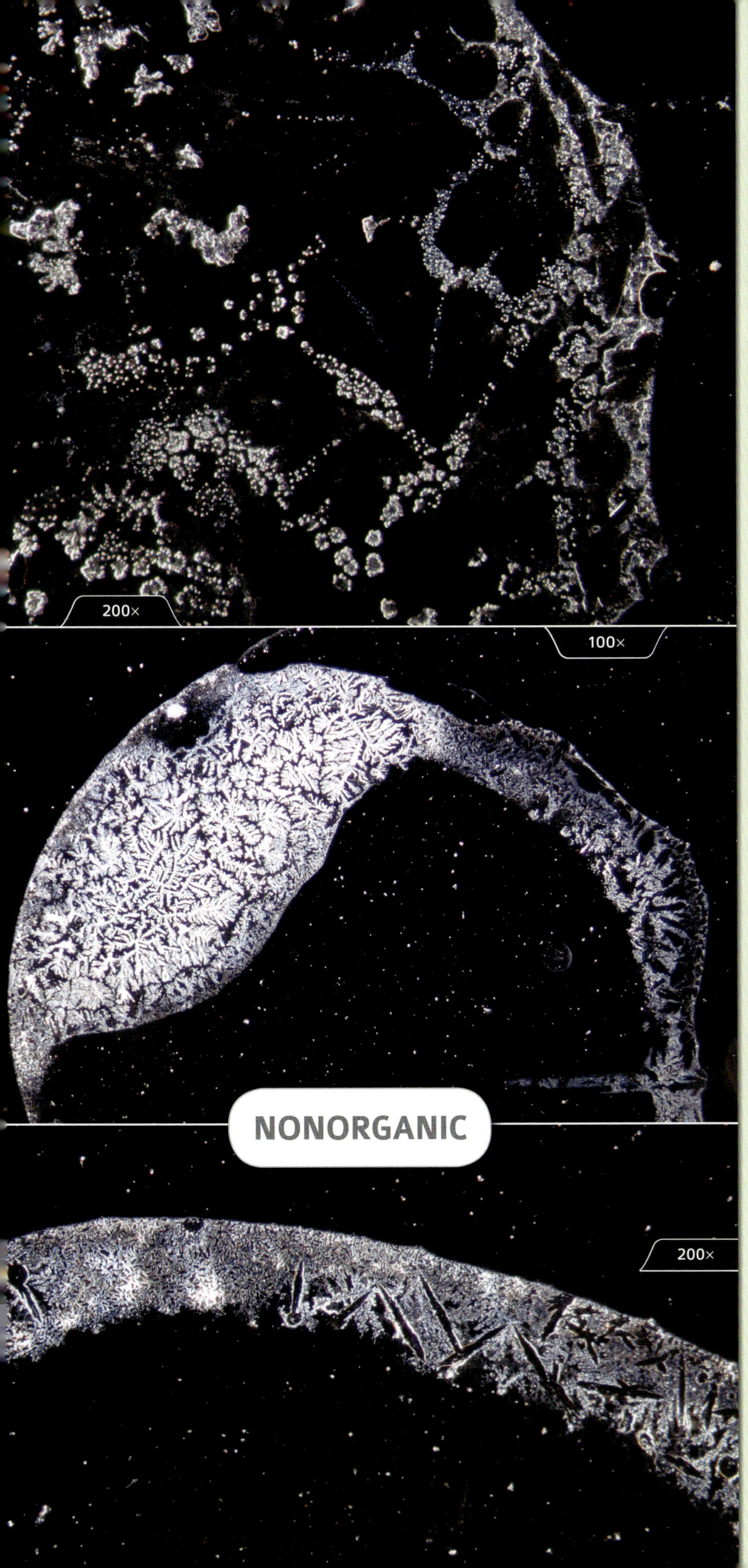
200×
100×
NONORGANIC
200×

100×
100×
ORGANIC
100×

NONORGANIC

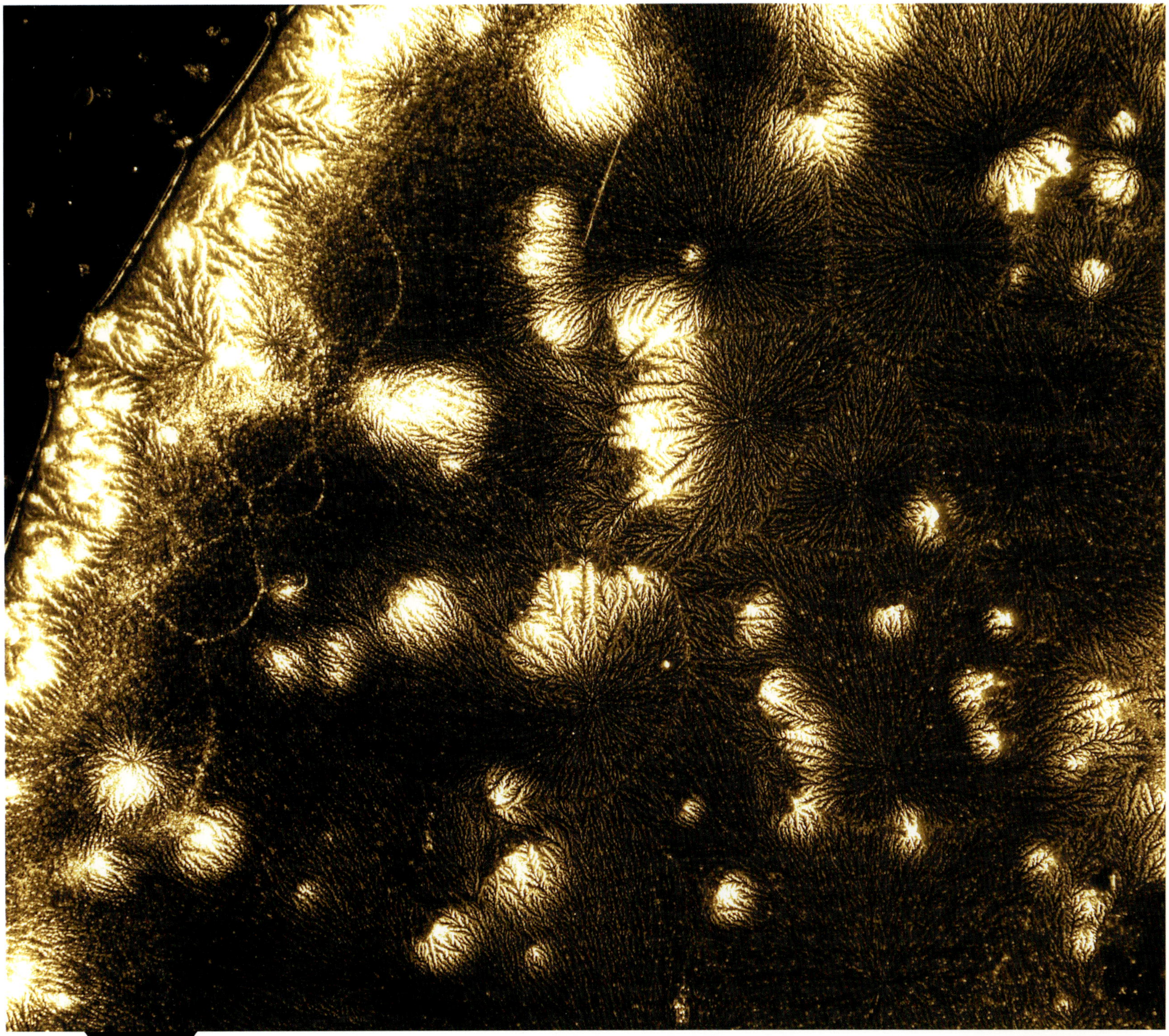

200×

ORGANIC

Millet

The organic millet shows us a powerful centre in a meadow strewn with "flowers-of-life." In the various images the organic millet reveals still other plant-like forms.

The nonorganic millet shocks us with a yawning emptiness. One area of plant-like forms is pressed to the sides. These are the only remains of the naturally beautiful and life-fulfilling harmony of this nutritious plant.

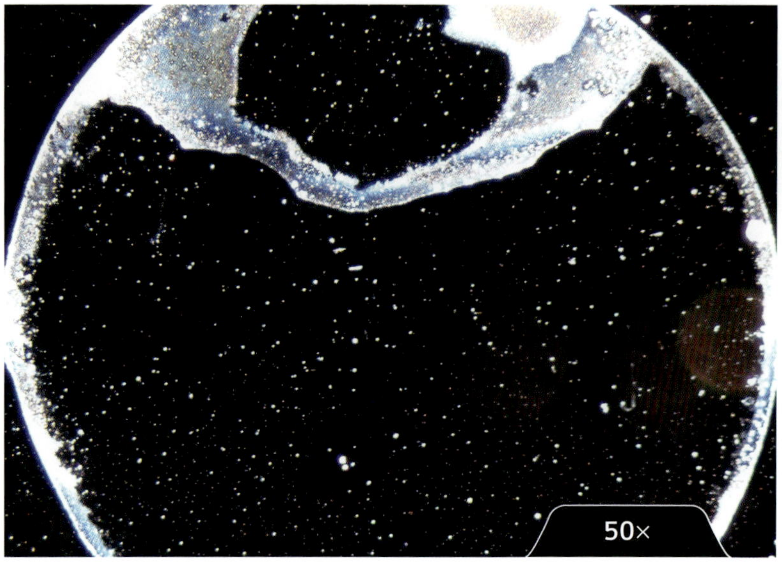

NONORGANIC

ORGANIC

200×
200×
200×
NONORGANIC
ORGANIC
100×
100×

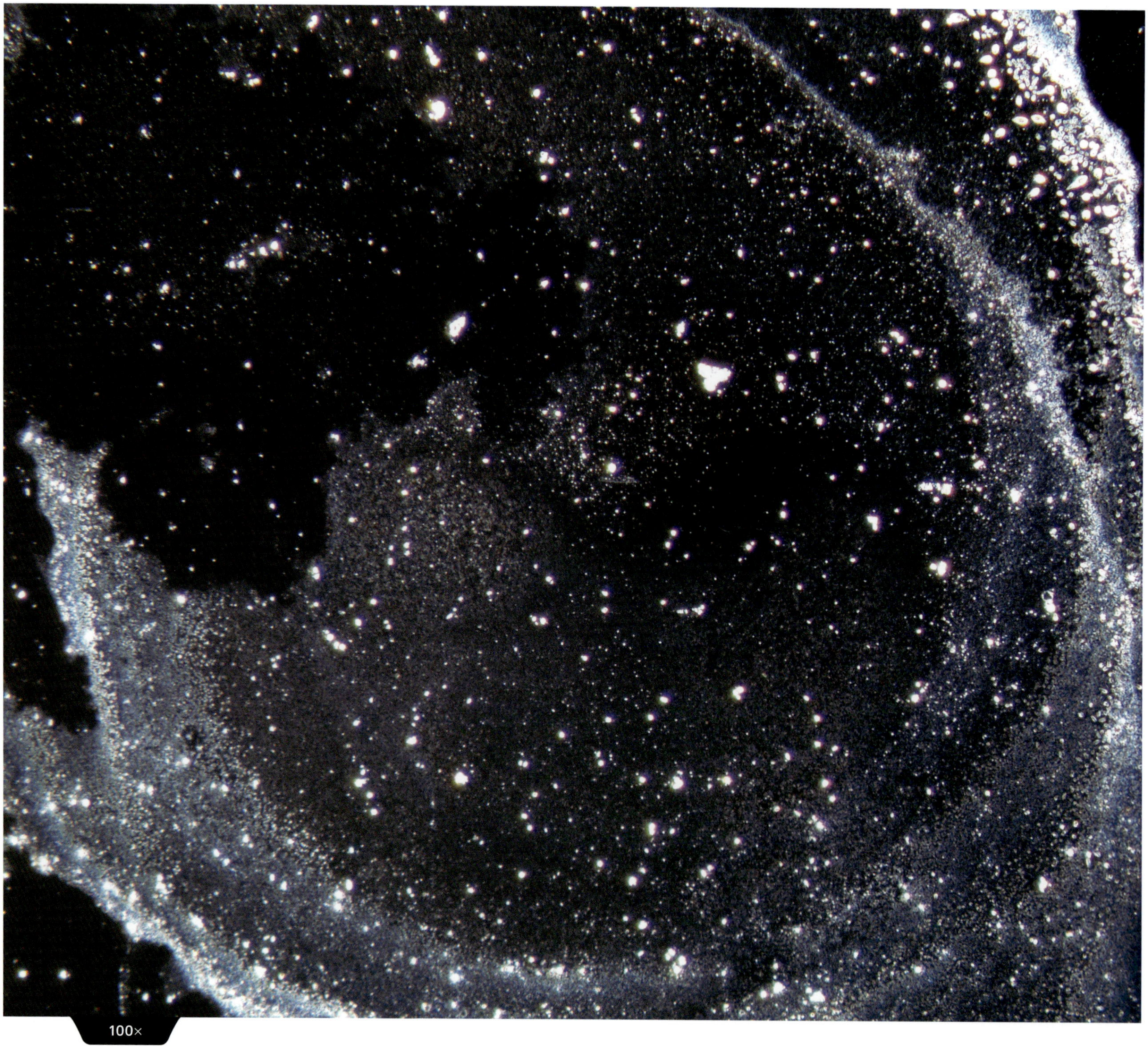

NONORGANIC

100×

ORGANIC

Buckwheat

The organic buckwheat is bursting with life-energy in beautiful and powerful plant-shapes, creating diverse landscapes that would be easy to thrive in.

The nonorganic buckwheat is hit hard. It is almost unimaginable, yet the pictures speak loud and clear: nothing from that tremendous wealth of life remains but a handful of dry desert dust.

NONORGANIC

ORGANIC

200×
100×
100×
NONORGANIC
ORGANIC
200×
100×
100×

NONORGANIC

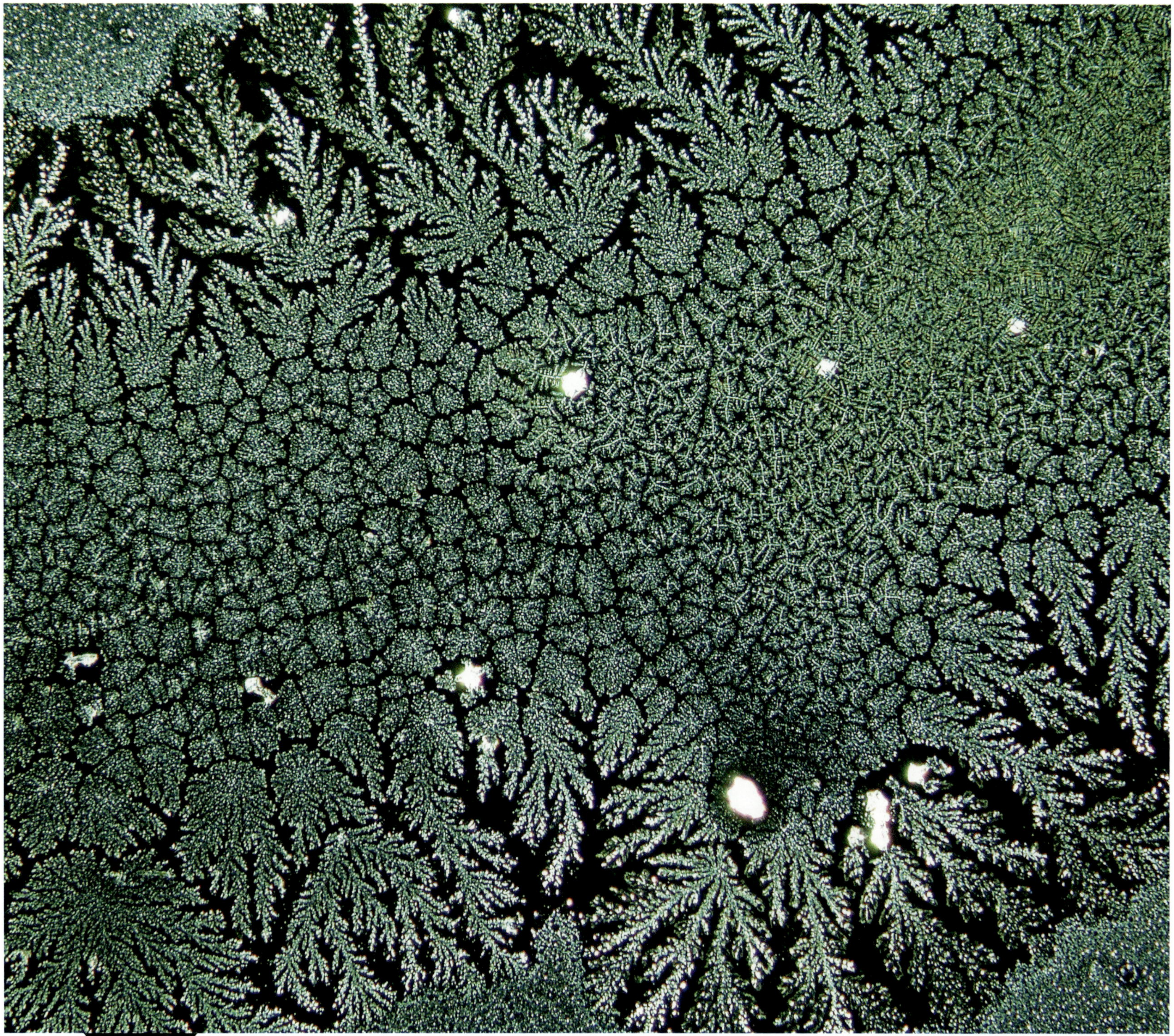
100×

ORGANIC

Corn

The organic corn naturally possesses an extremely powerful essence that shines like the sun, much like the organic sunflowers. The life design principle in the organic corn passes on its impulse in concentric circles, with each circle in turn creating new formations. At a glance we encounter amazingly organized sun-power-landscapes. In the more detailed images we find ourselves amidst mighty yet graceful power currents in constant flow. The pictures remind us somewhat of perfectly organized fireworks, yet even the most masterly fireworks technician will not be able to achieve this degree of subtlety and variety, with blossoming star-images lighting up the night sky. After all, ordinary fireworks pale in a few fleeting moments, whereas the sun-power in organic food continuously nourishes us.

The images of the nonorganic corn reveal to us the dramatic and tragic battle of the natural order forces, which have been badly crushed by the chemical forces. At best the natural forces are pushed into reservation areas along the edges, like natives inside confined territory, where they are still allowed to express the beauty and wealth of their culture, as long as they can muster the strength to do so.

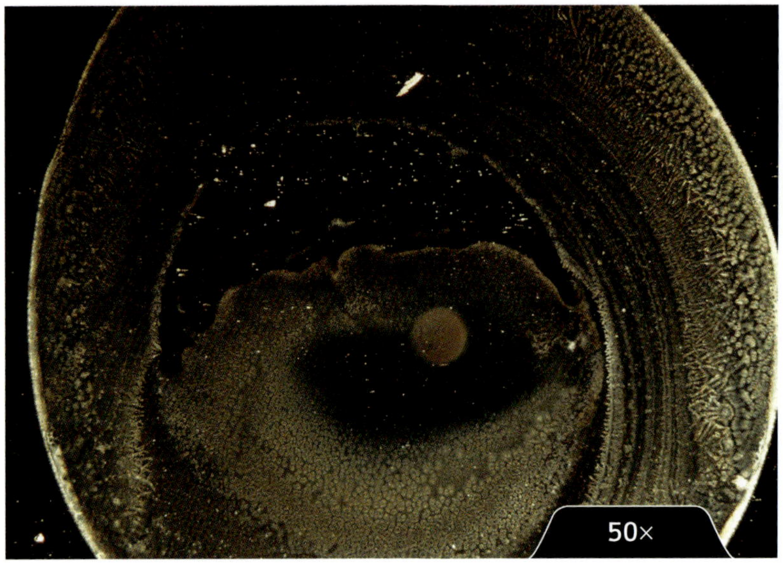

NONORGANIC

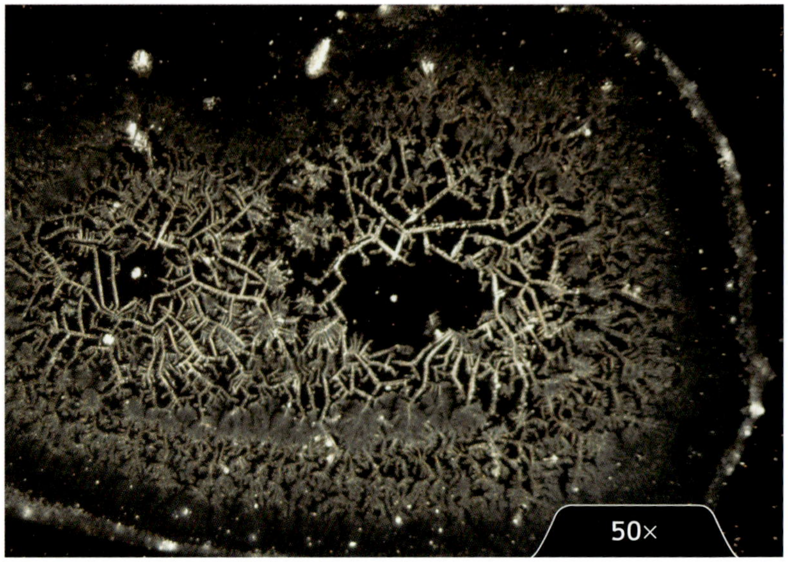

ORGANIC

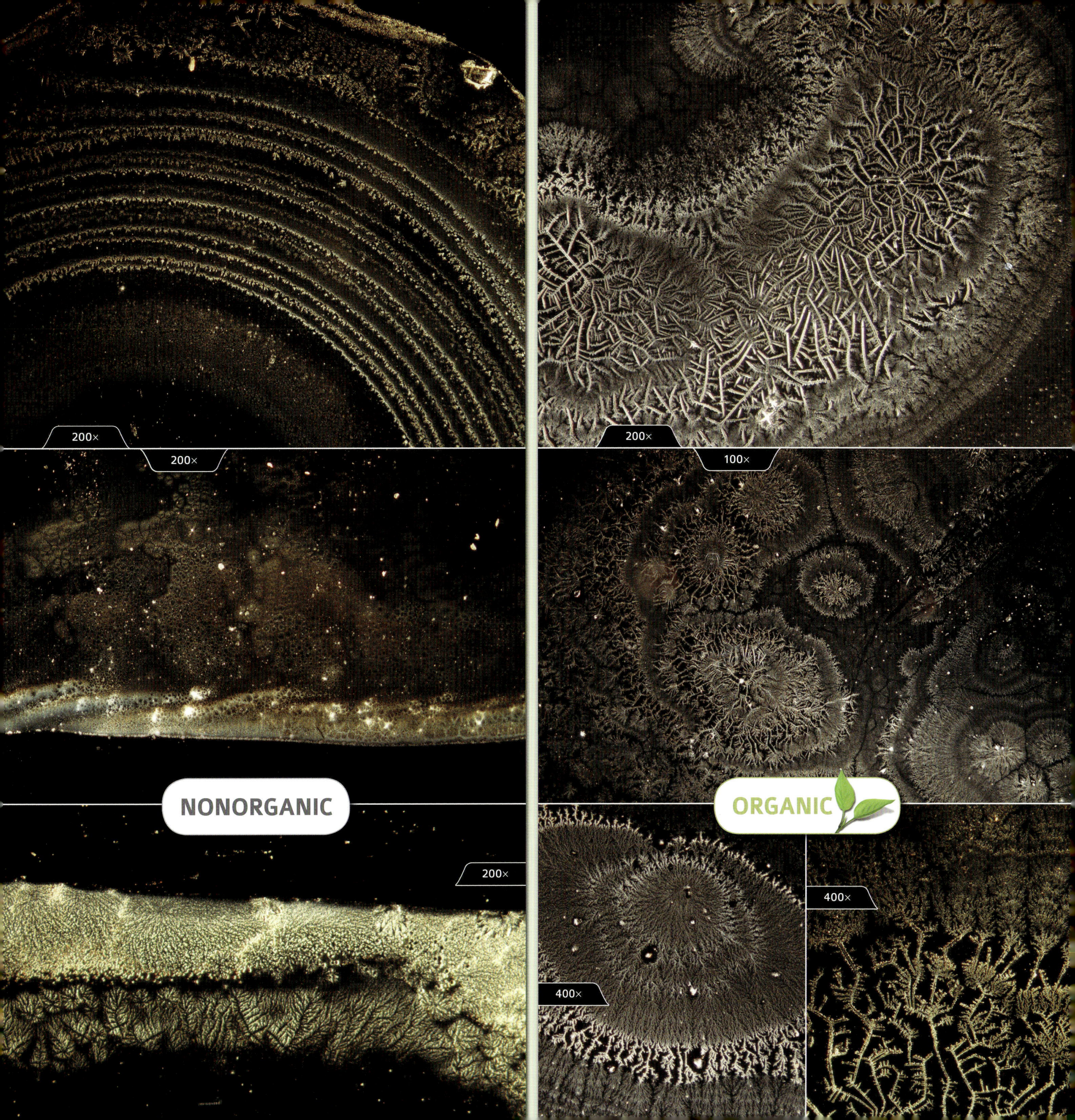
200×
200×
200×
100×
200×
NONORGANIC
ORGANIC
400×
400×

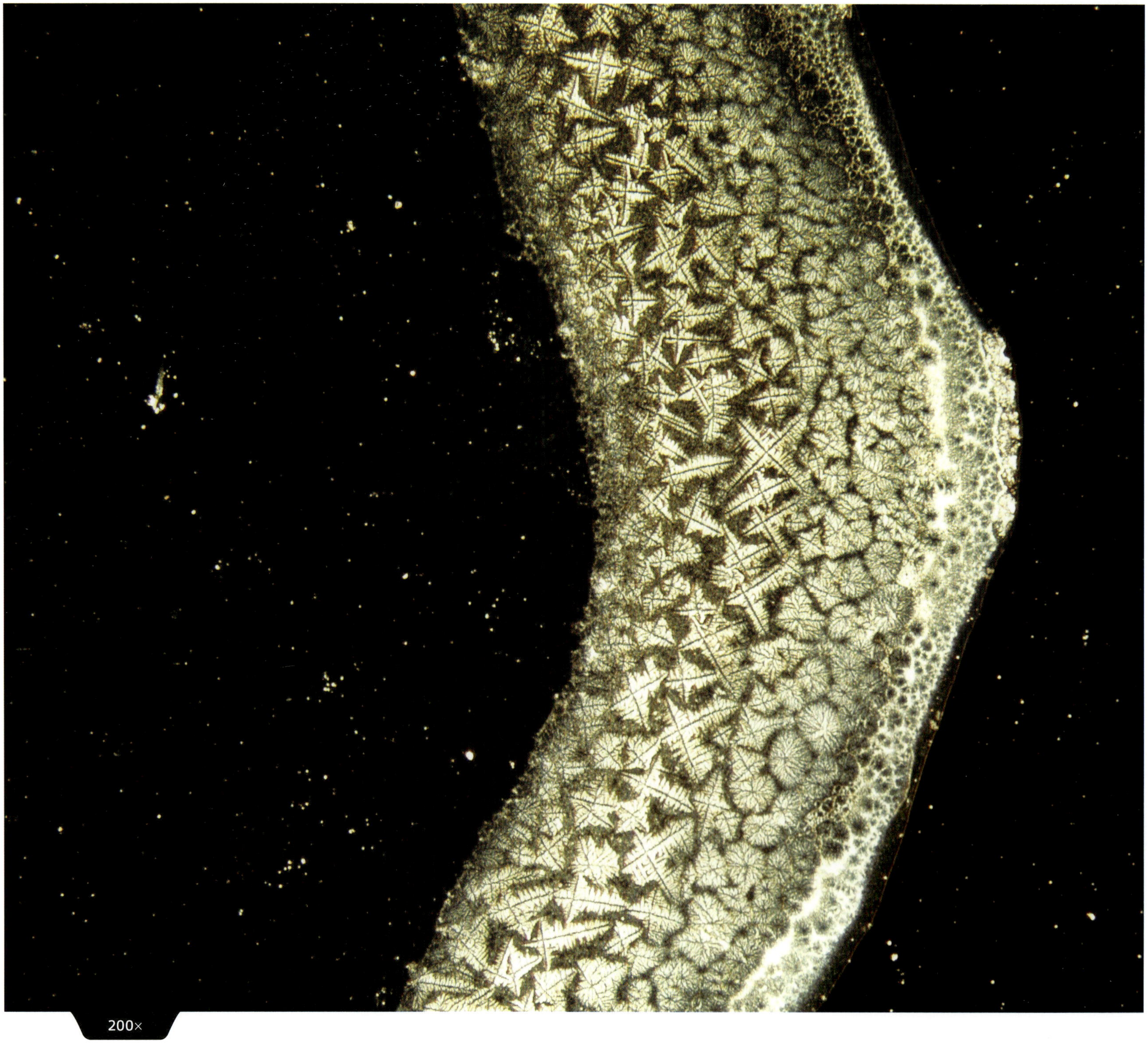

NONORGANIC

ORGANIC

Almonds

By nature the organic almonds are wonderfully balanced, which is beautifully revealed in their life-energy. On the one hand, they roll out a carpet of black cut-outs of "flowers-of-life," and on the other hand they combine these flowers with the images of their fruits in a most artistic manner: a natural work of art, where a fully developed almond tree can be found in every part of the almond.

Compared to the other nonorganic foods the life-energy of the nonorganic almonds is much better preserved, showing fully informed droplets and the occasional beautiful formation, possibly because they have been subjected to significantly fewer chemical substances. Nevertheless, most structures have lost their natural coherence.

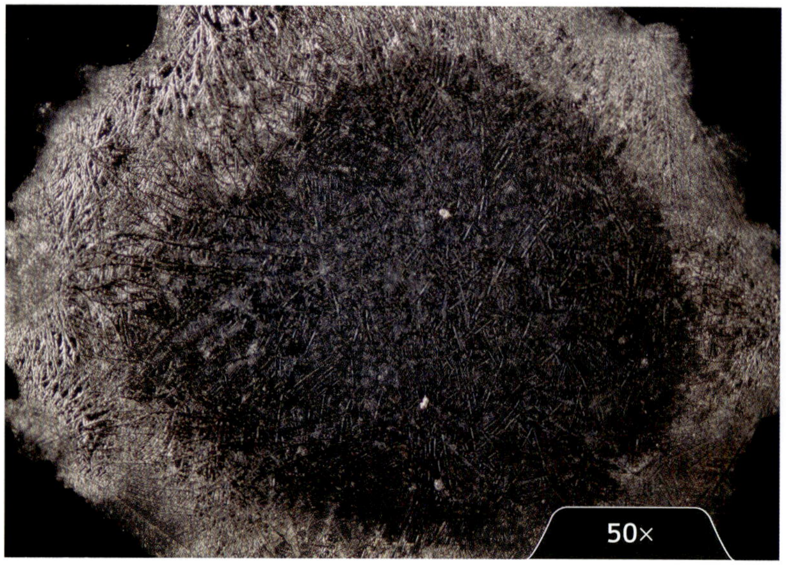

NONORGANIC

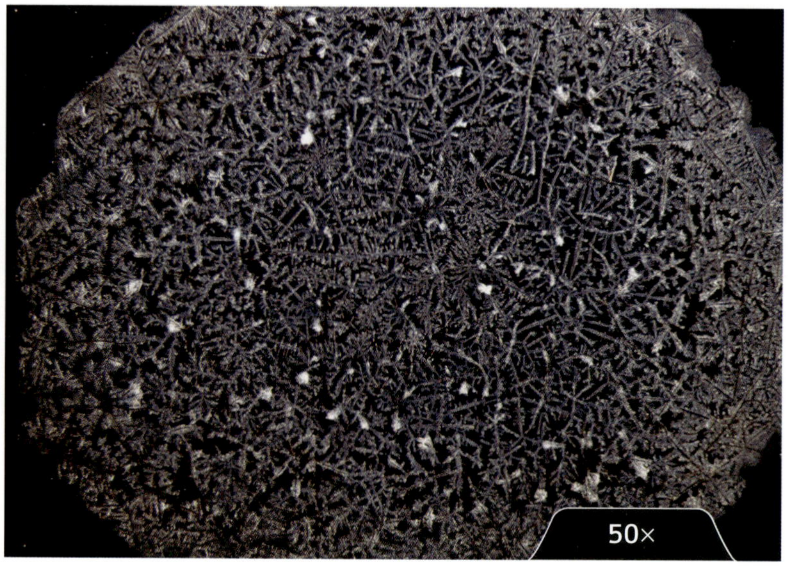

ORGANIC

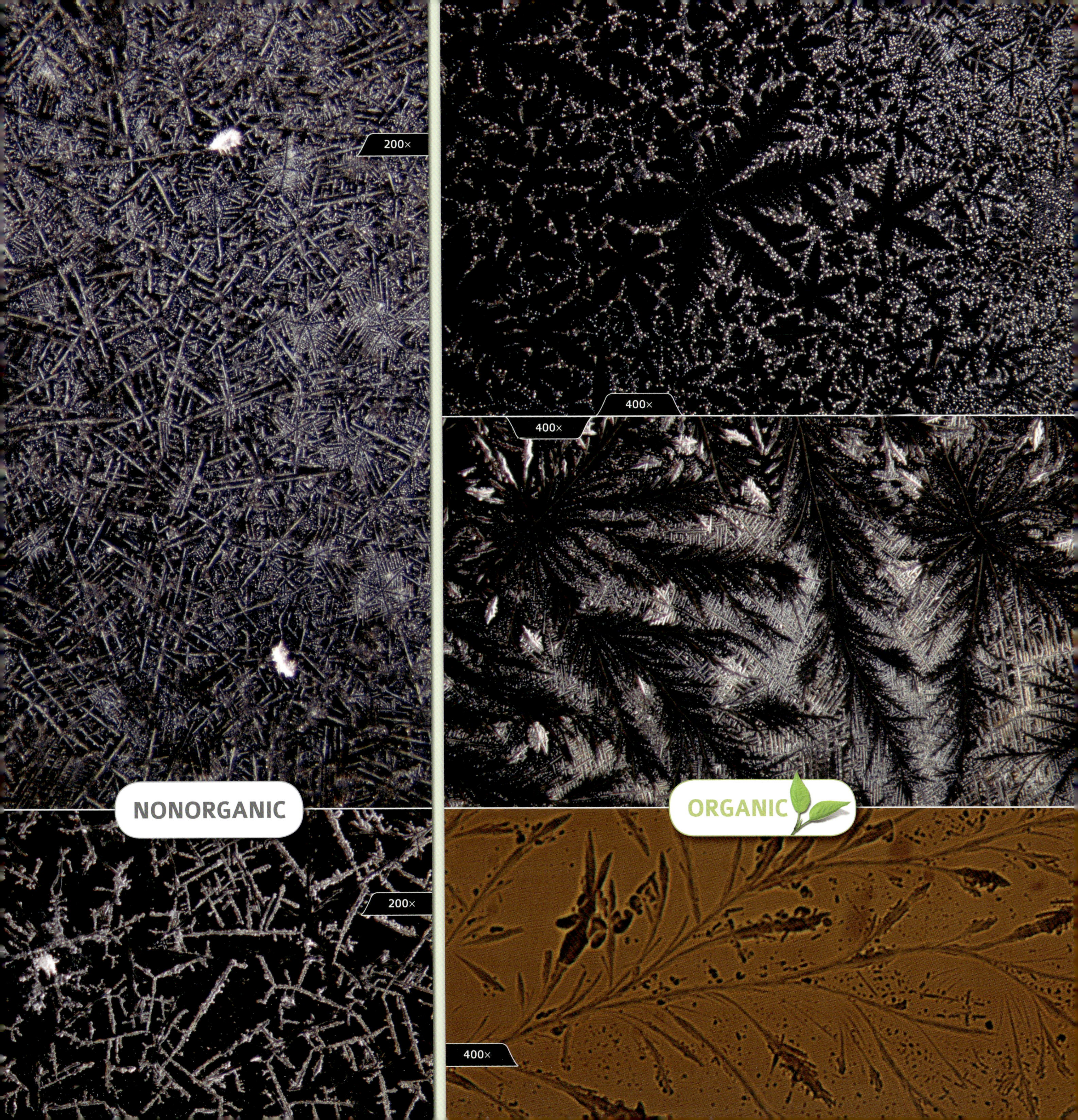
200×
400×
400×
NONORGANIC
ORGANIC
200×
400×

400×

NONORGANIC

400×

ORGANIC

Cashews

The organic cashews reveal themselves to us with a characteristic and powerful appearance: a wide border of large and densely growing six-petalled stars surrounds a neatly arranged field of parallel lines. In one example these grow together like ladders with organic rungs; elsewhere they are filled with different kinds of plant formations growing closely together. Where they meet in the middle, four-edged stars can be found. Is it not fascinating how the Creator always manages to find new forms?

The nonorganic cashews are dominated, we might say, by aggressive forces, pushing all natural forms harshly to the sides or completely destroying them. The remnants are beautiful, but they are unable to spread.

NONORGANIC

ORGANIC

200×
50×
100×
NONORGANIC
50×
200×
400×
ORGANIC
200×

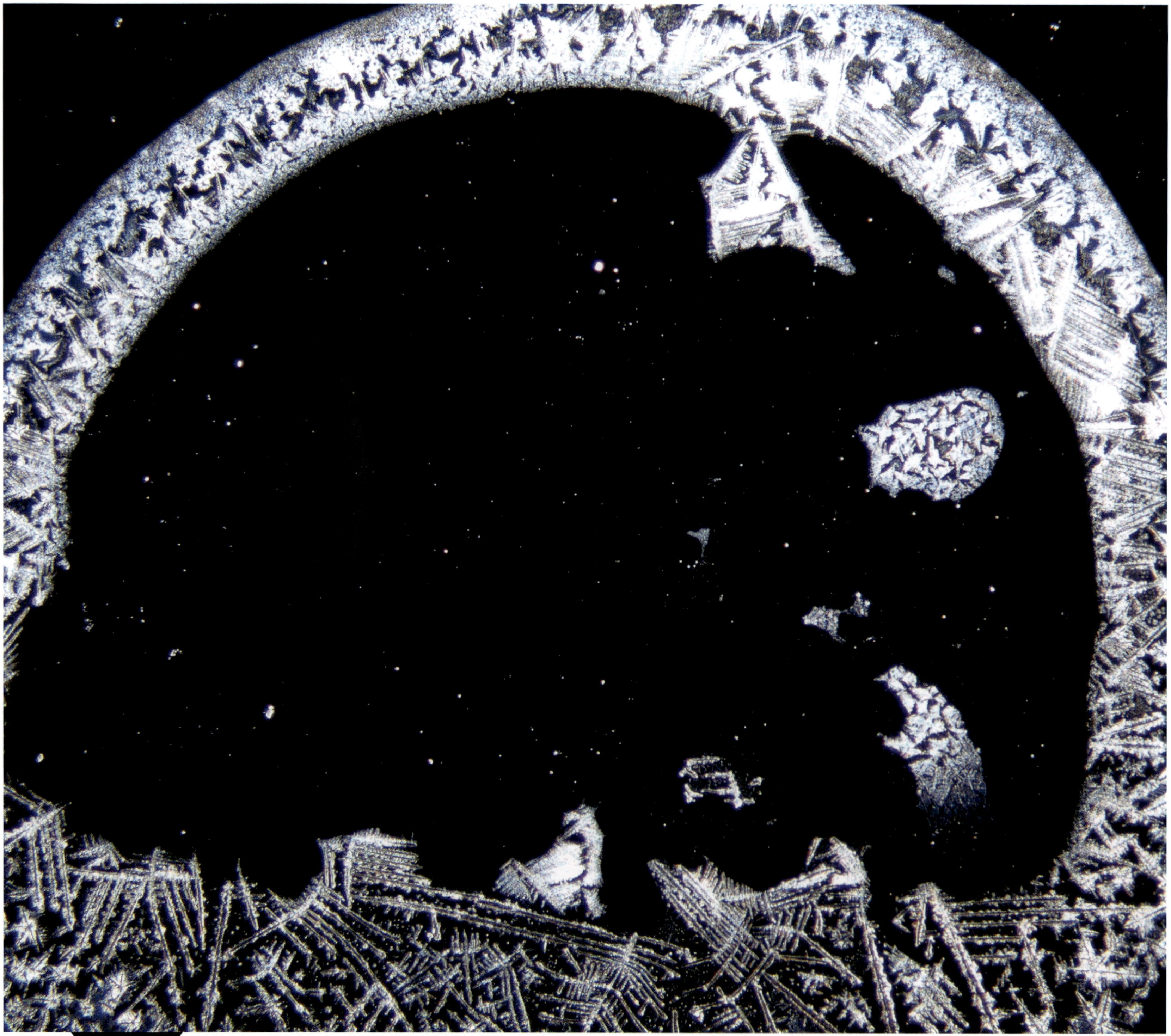

100×

NONORGANIC

200×

ORGANIC

Hazelnuts

The life-energy of organic hazelnuts is saturated with fine plants, unfurling like a river and growing quite powerfully. In addition, their blossoms can be seen in many images, forming rich impressions of flowing life-currents.

Unfortunately the nonorganic hazelnuts are demolished into barren deserts, with remnants of plant-like fragments sticking out of the desert sand. Fully preserved plant forms are only found along the edges, but quite rarely. It is clear that the natural forces wanted to bring to the fore beautiful shapes and forms, but were unable to. What happened here?

NONORGANIC

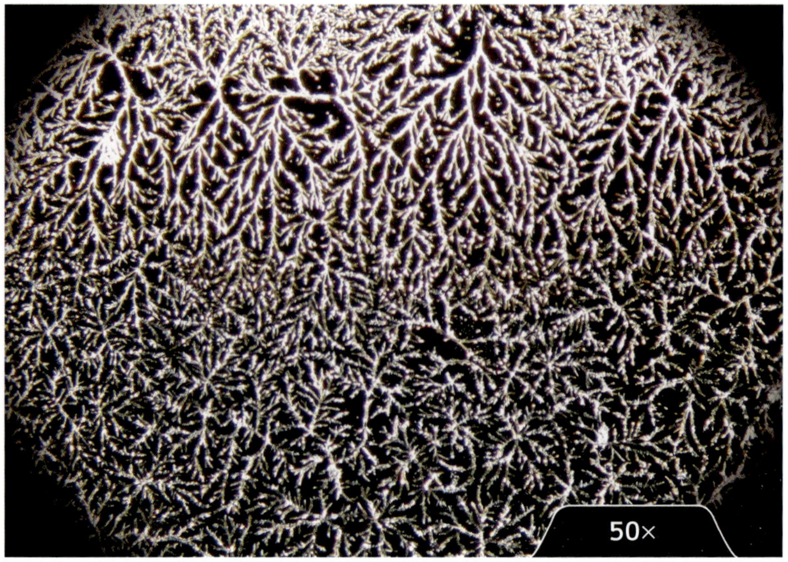

ORGANIC

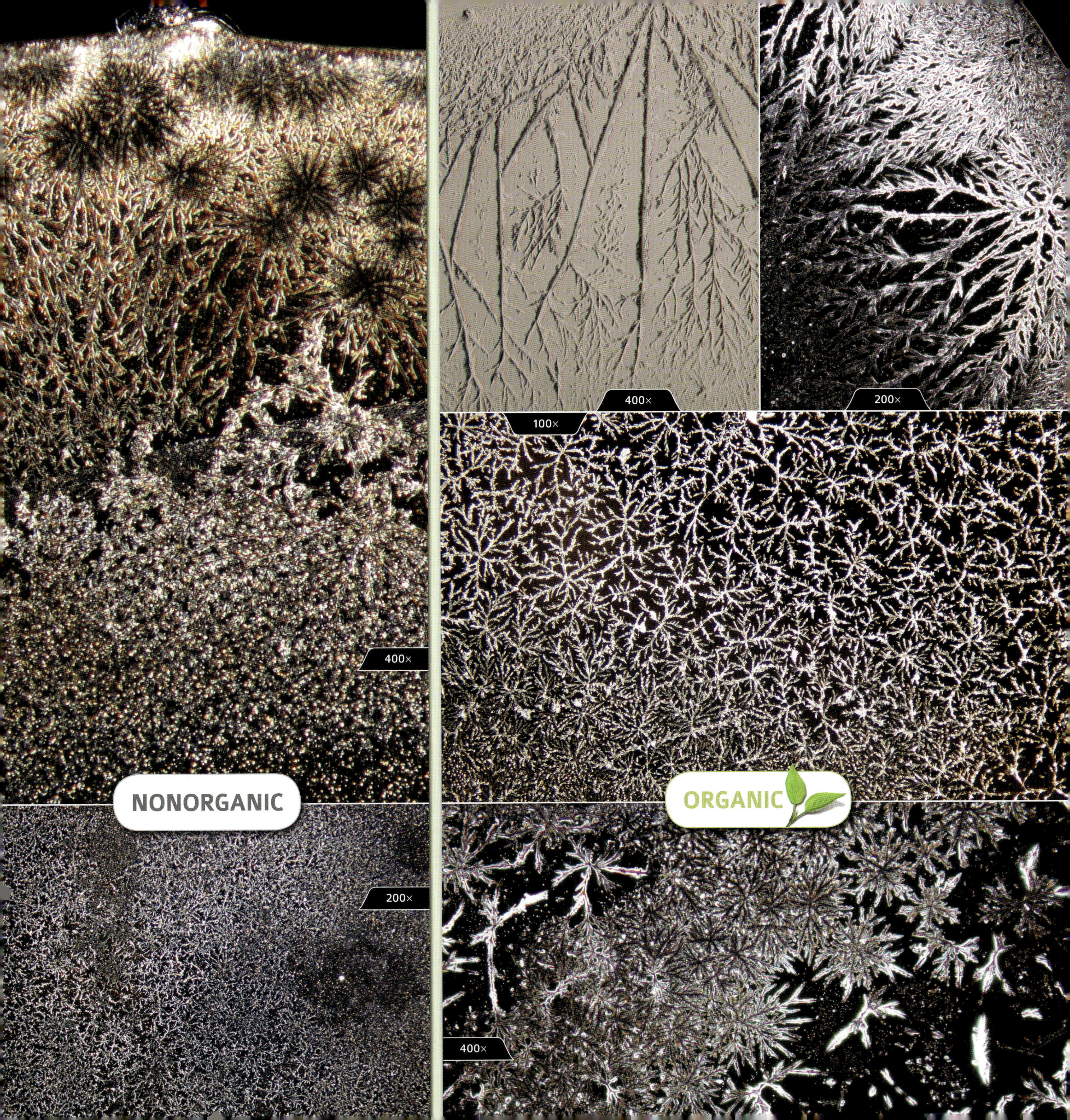
400×
200×
100×
400×
200×
NONORGANIC
200×
ORGANIC
400×

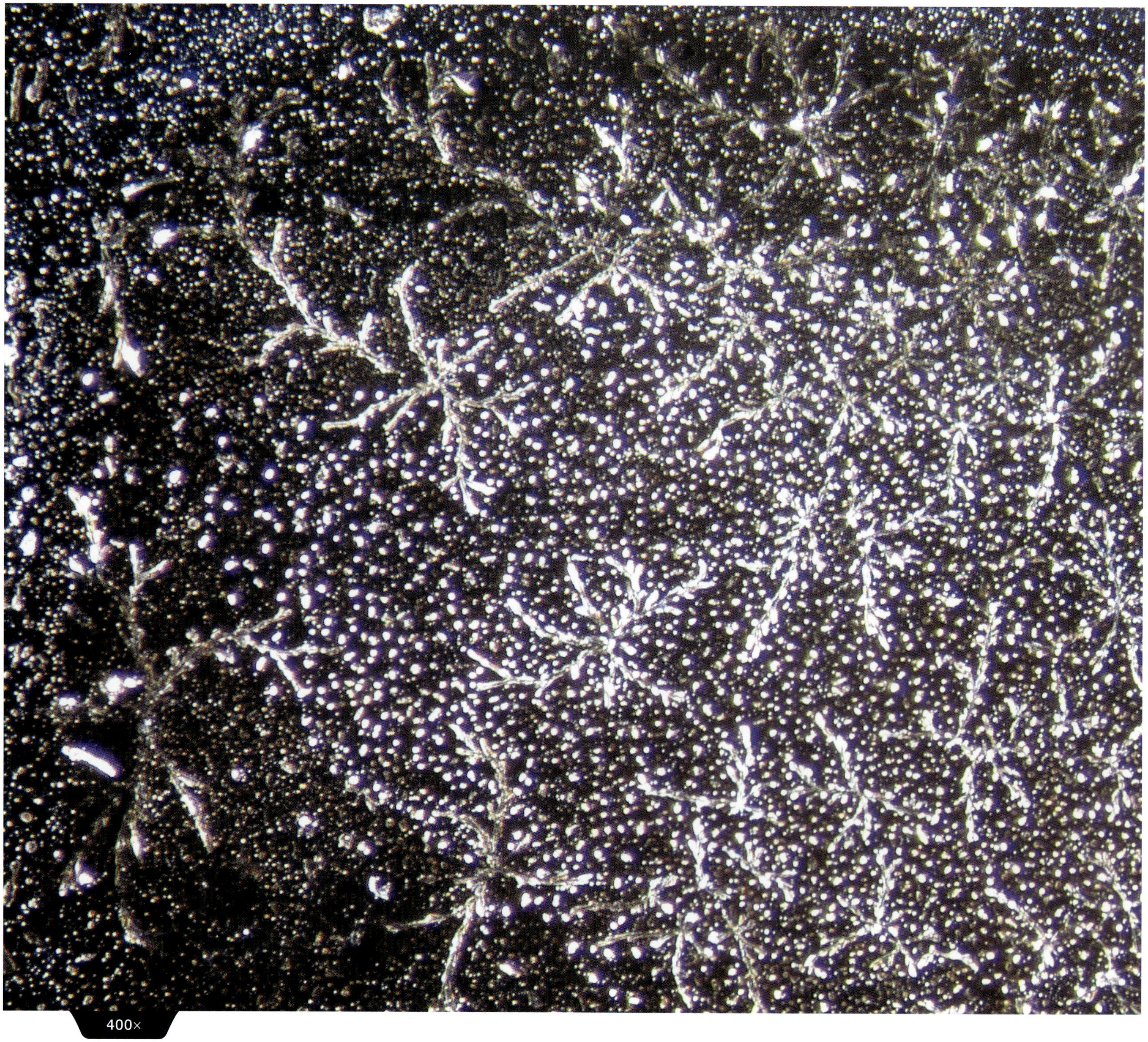

NONORGANIC

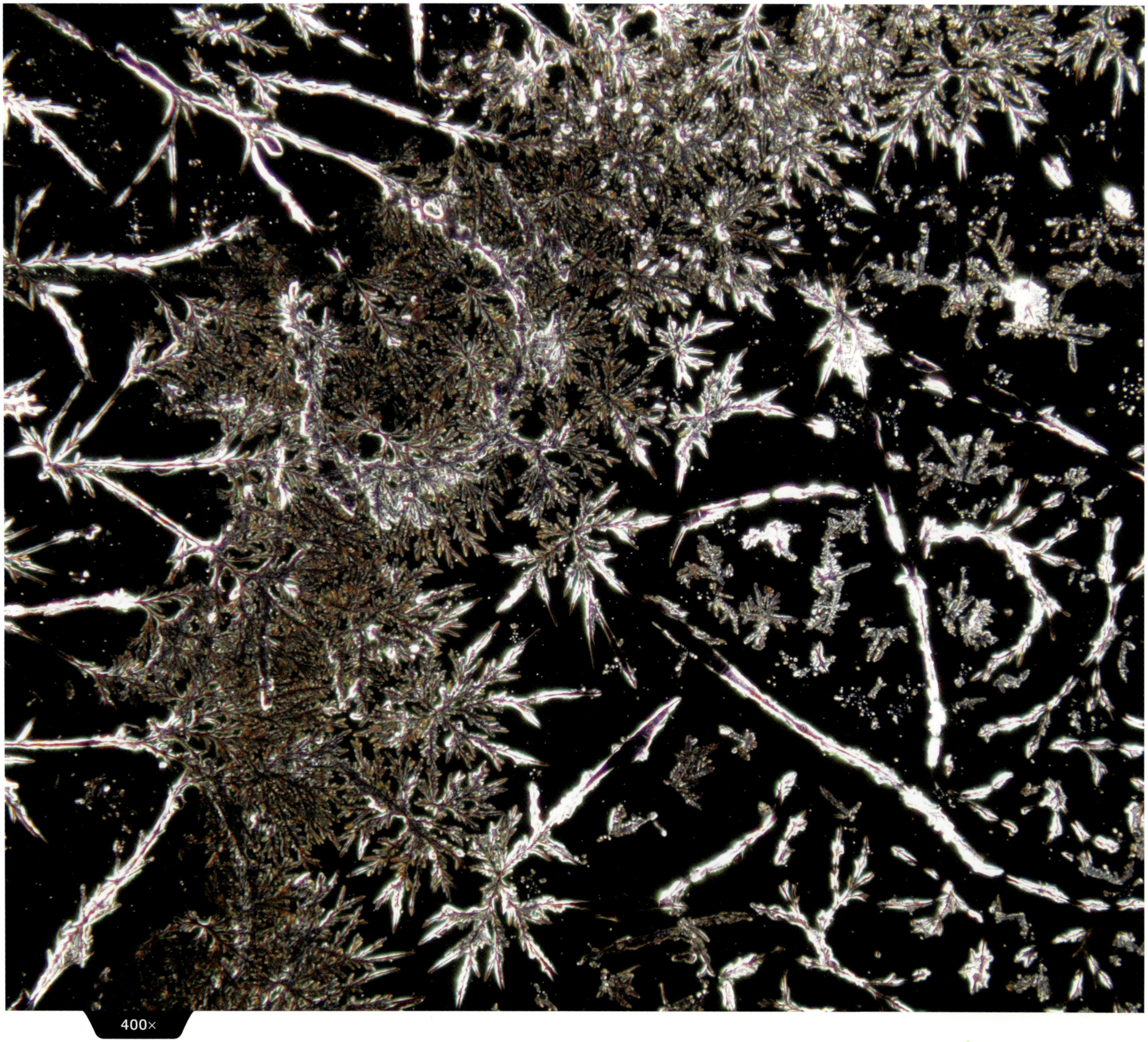

ORGANIC

Sunflower oil

Organic sunflower oil has obviously received its order force (life design principle) from the sun, having absorbed its essence. The sun shines so brightly in these images, that either the whole droplet is shot through with impulses arising from the centre—often reaching out over seven layers or horizons—or a series of cells is formed with a powerful sun shining in each one. This gives birth to a veritable empire of the sun. The sunflower does its name justice wonderfully—but only the organic one!

In the nonorganic sunflower the order force (life design principle) is not worth the name anymore, as there is hardly any order or strength to be found there. The awesome natural information is pulverized and sedimented, leaving only some hieroglyphic fragments. The shocked observer may well ask: what kind of force is behind this, that can manage to annihilate the total power of the sun? It would have to be tremendous, but so destructive that this food cannot remember anything. It must be a force that eliminates the natural information. Do we really want to take inside us such a force through our food, and on a daily basis?

NONORGANIC

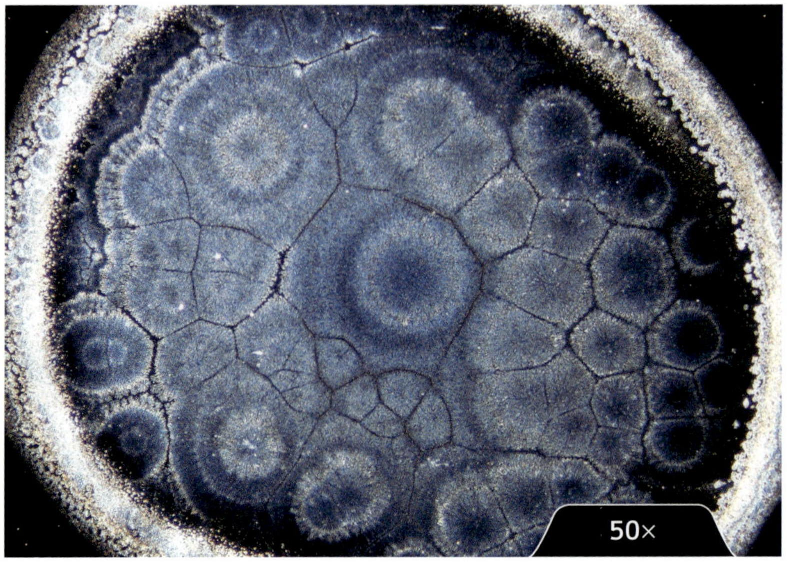

ORGANIC

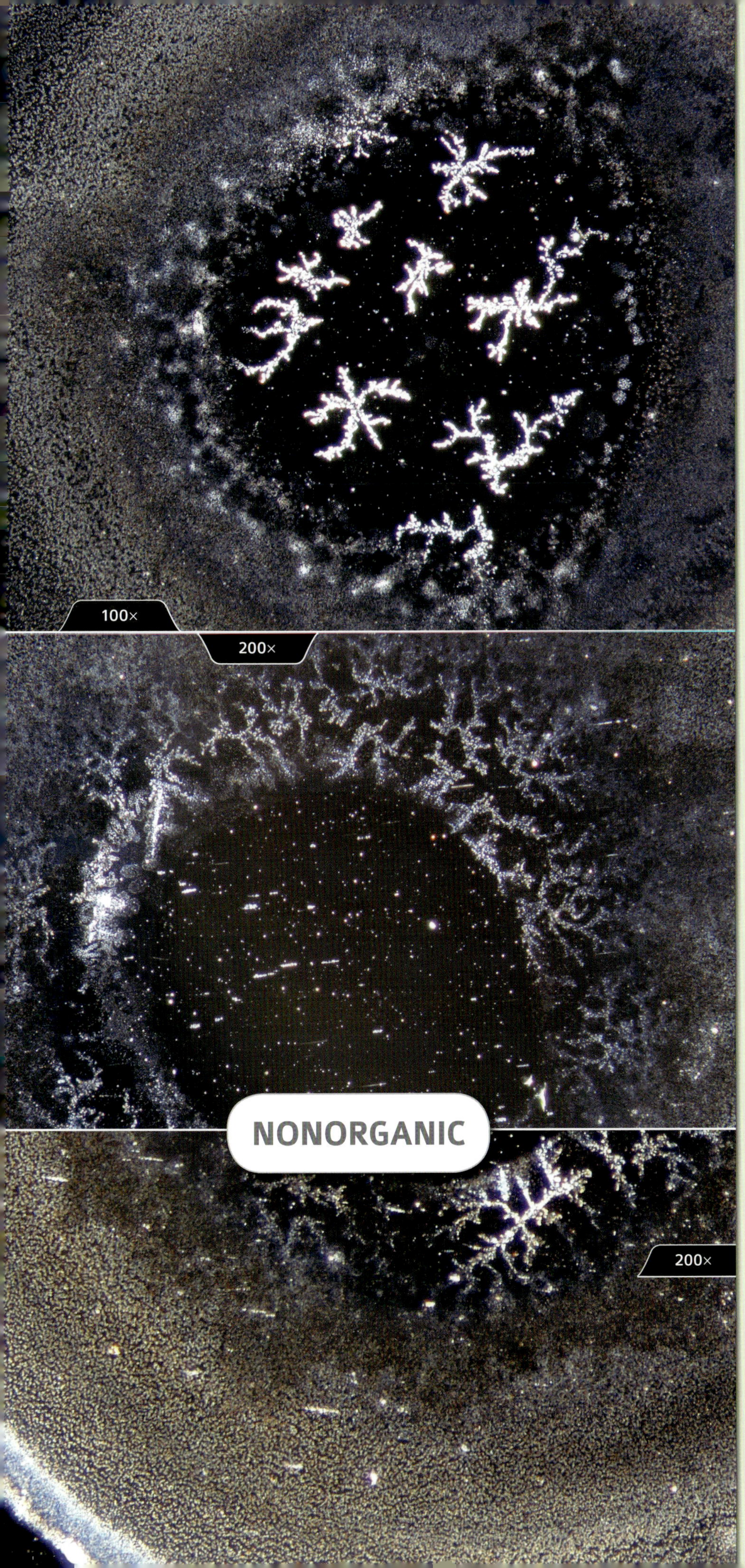
100×
200×
NONORGANIC
200×

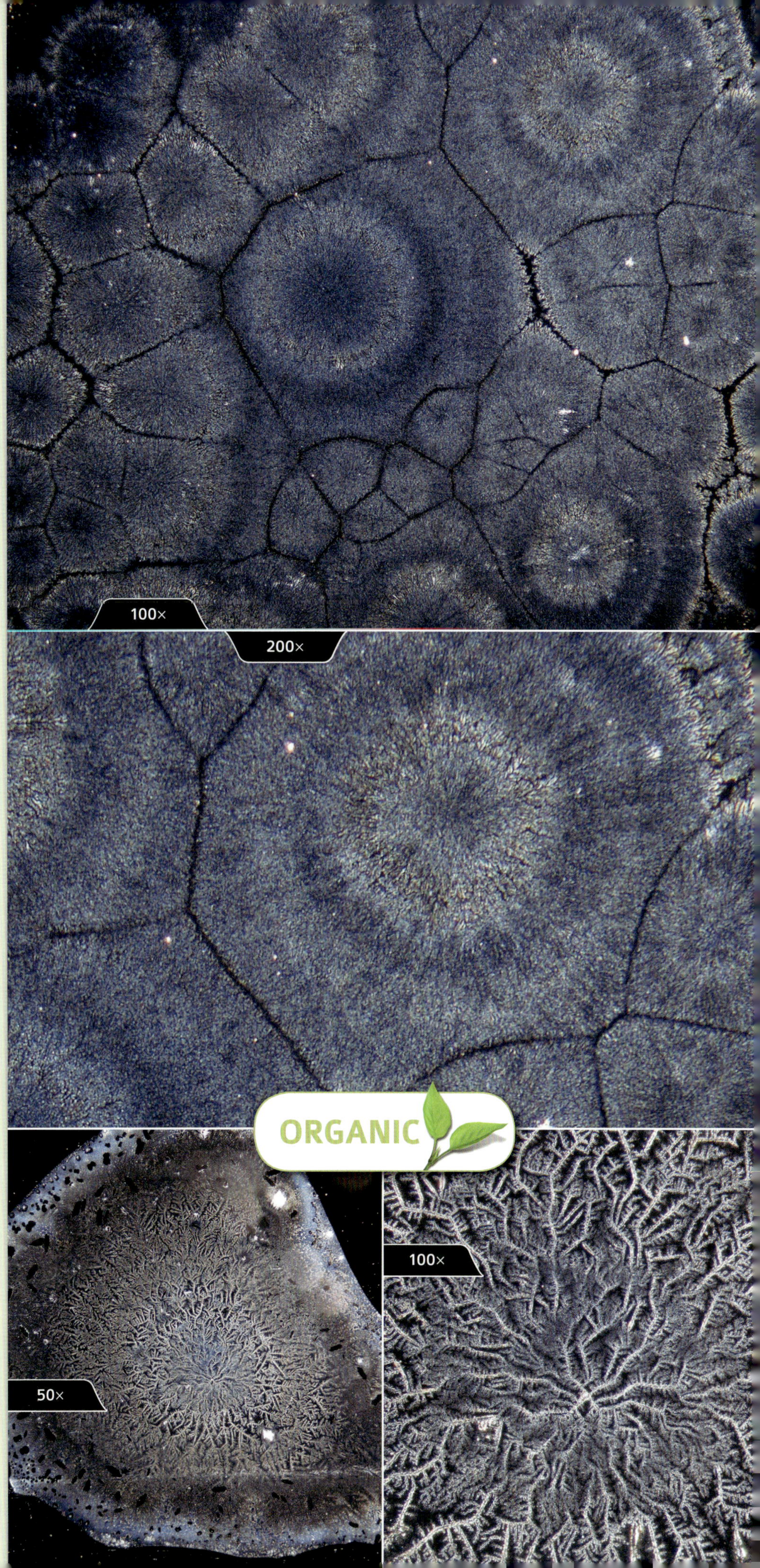
100×
200×
ORGANIC
100×
50×

100×

NONORGANIC

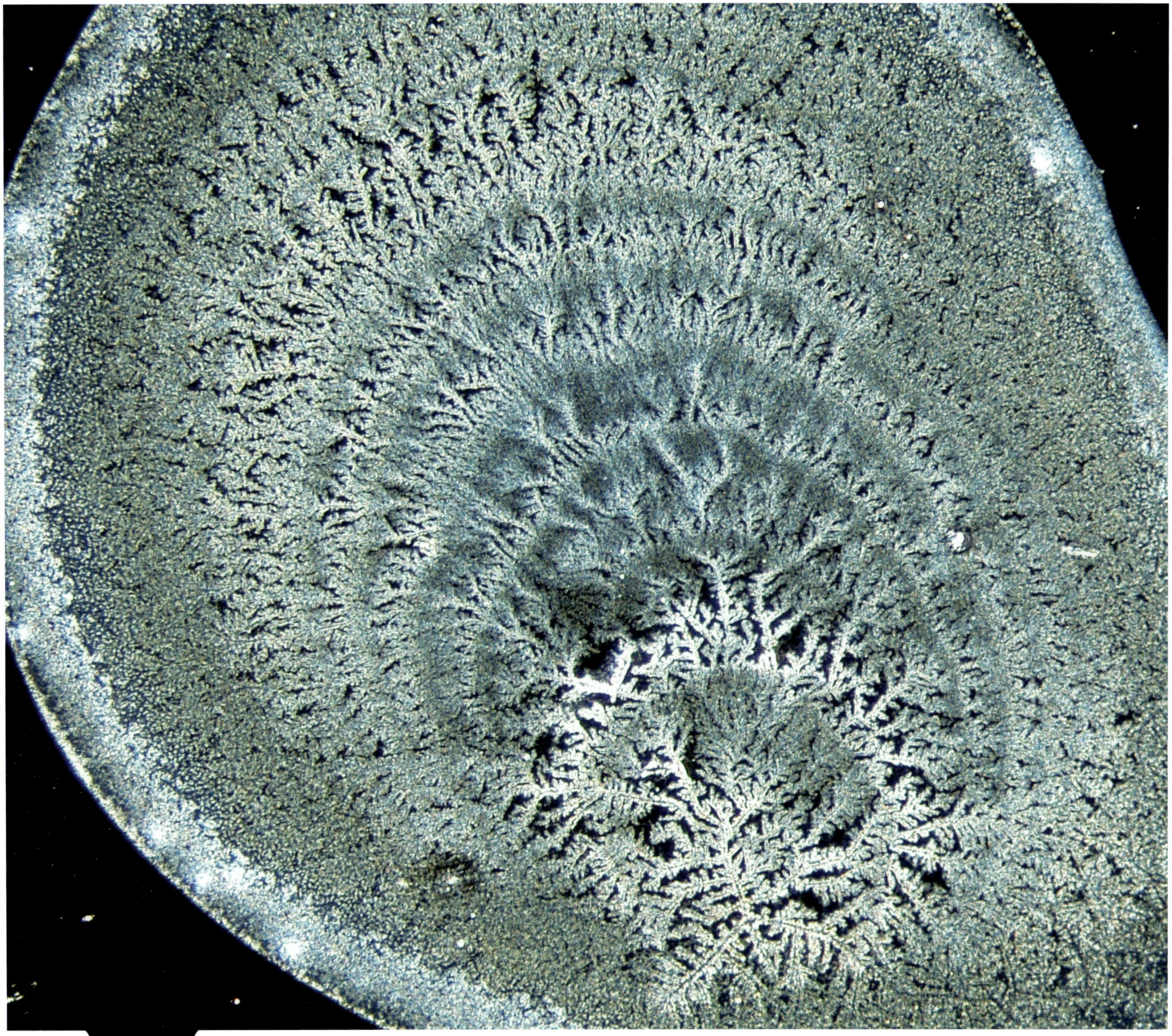

100×

ORGANIC

HighOleic
Organic sunflower oil

In the 1980s high oleic sunflowers were cultivated as a new variety through crossbreeding of selected sunflower plants, with the idea of creating a sunflower oil with superior nutritional qualities for withstanding better the heat. When oils are heated, the content of mono- and polyunsaturated fatty acids plays a central role, as monounsaturated fats can withstand oxidation and heating much better than polyunsaturated fatty acids. During the heating process unsaturated fatty acids are split and polyunsaturated fatty acids form substances that may turn toxic in the organism, such as trans fatty acids, free radicals and even carcinogenic substances.

Sunflower oil contains about 23% mono- and 60% polyunsaturated fatty acids. Because of the high content of polyunsaturated fatty acids, it is better to avoid heating sunflower oil, and to use it in its wholesome cold condition in cold dishes like salads.

Through careful cultivation in the late eighties it became possible to create an organic sunflower oil with about 80% mono- and 13% polyunsaturated fatty acids. This oil is similar to olive oil, whose content of 71% mono- and 10% polyunsaturated fatty acids makes it suitable for hot dishes as well.

Therefore since the beginning of the 1990s, Soyana has only used high oleic organic sunflower oil for all its heated foods, like Swiss Drinks (milk alternatives), tofu burgers and sausage and meat alternatives. In line with Soyana Quality Principle 4 ("paying attention to 1001 details") this was done in order to offer customers the best possible food quality.

For this reason we were especially interested in the life-energy of high oleic organic sunflower oil. To our surprise and joy our investigations yielded amazingly beautiful images, which we would like to share with you here.

High oleic organic sunflower oil, which Soyana uses in its foods and sells in glass bottles, possesses a life-energy that shows an even more beautiful life design principle than organic sunflower oil. These unique images are a testimony to the solar powers operating in it. Breathtaking landscapes arise, composed entirely from "solar cells", manifesting giant, multilayered plant-artworks of a superior order. Whereas the solar impulses in the simple organic sunflower oil were able to spread out over seven layers or horizons, the high oleic organic sunflower oil operates on an even higher level of order. In the high oleic version two or more "solar cells" either join together to spread out in a single wave, or detach themselves from a tapestry of "solar cells", like droplets coming loose, increasing their power and building a new centre, like a hilltop from which new horizons or additional power sources stream forth. We can observe that each single "solar cell" has the potential to develop itself into an additional "central sun:" a universal principle of creation beautifully illustrated here. And how amazing it is to see that, while one "central sun" can often be found in the middle of a cluster, there may also be two, three or many! Furthermore, many of these artworks are suspended in a sea of "solar cells". One of these fascinating images adorns the cover of this book, showing the immense power that emanates from a flower of life, with its six characteristic flower petals.

We can see how this beautiful natural harmony arises from a supernatural divine order. When we are able to feel this, we forget our name and our role. Faced with such deeper and higher beauty, we are transformed into amazed and happy beings.

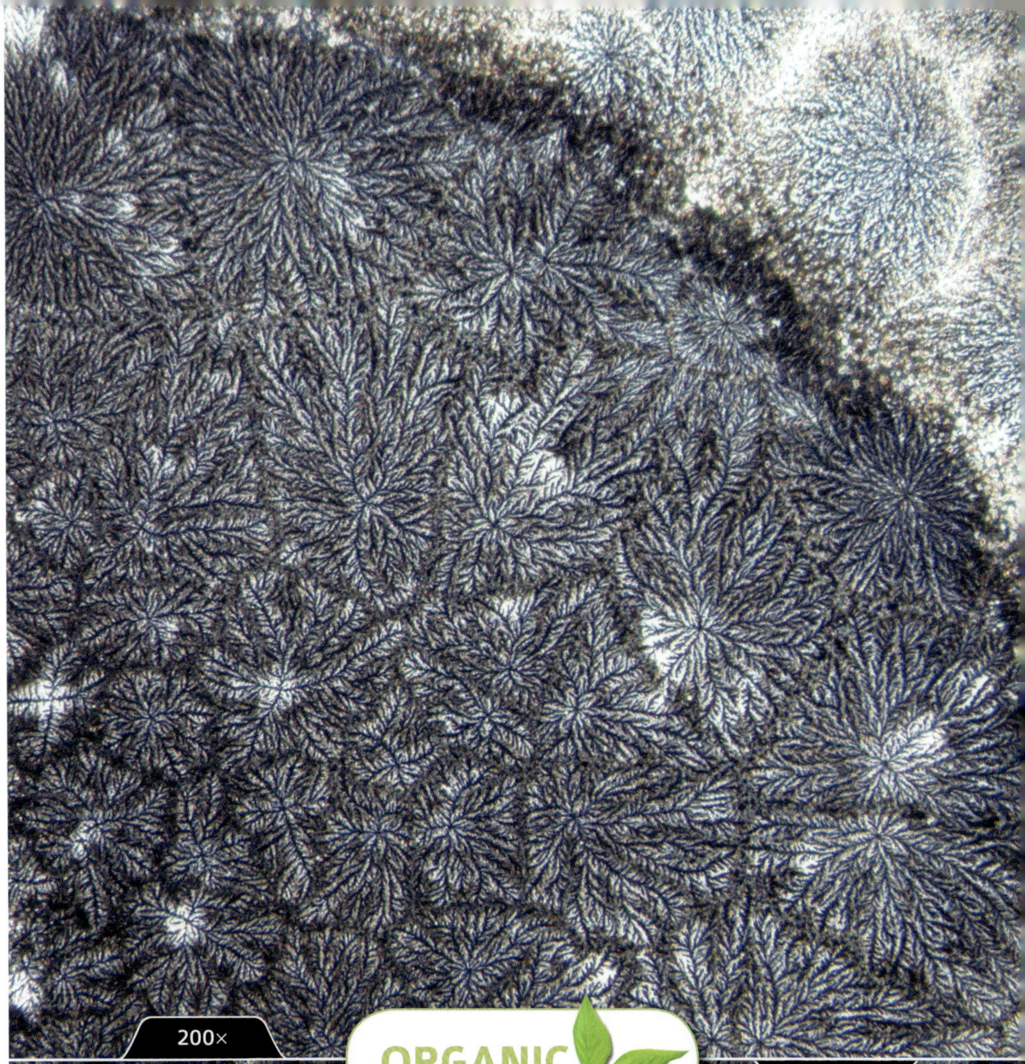

200×

200×
ORGANIC

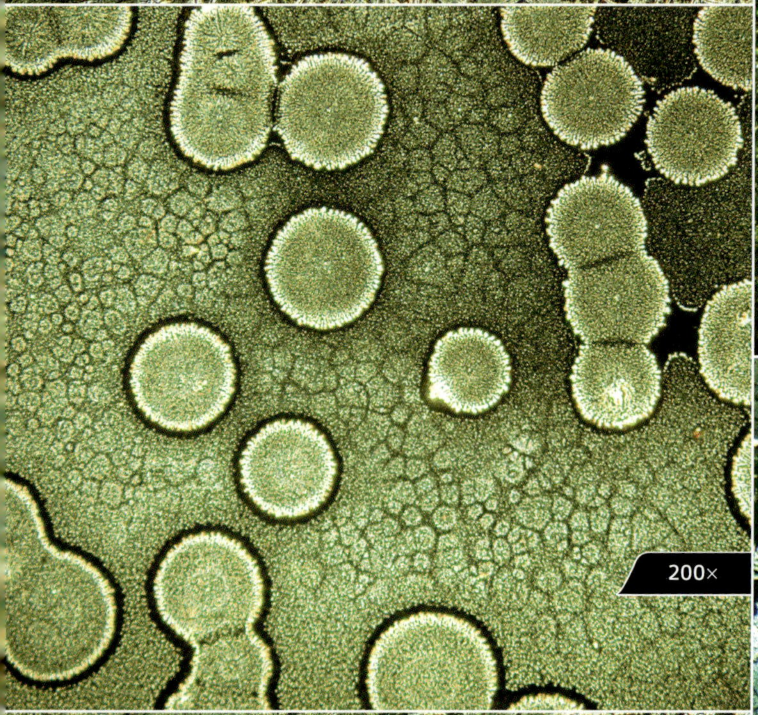
200×

400×

200×

200×
ORGANIC
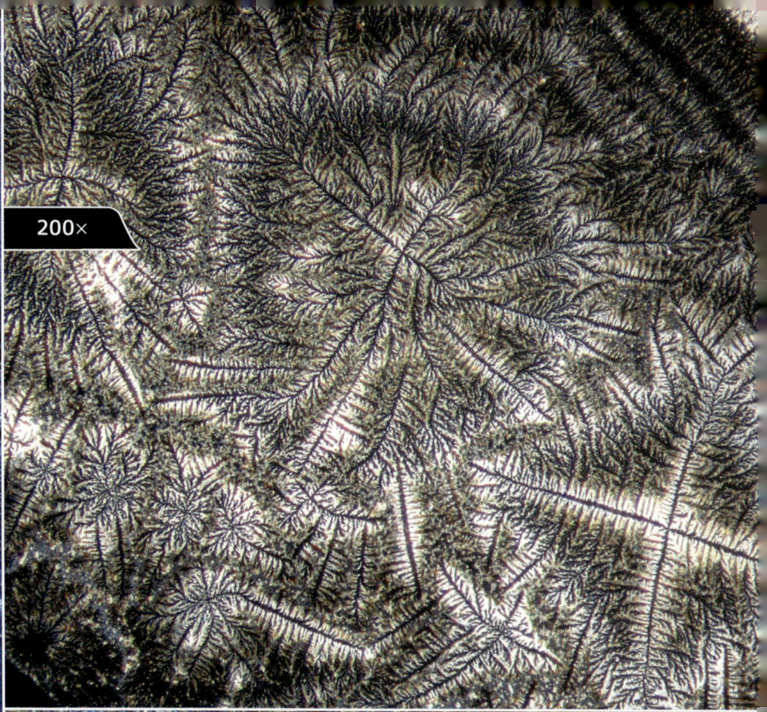
200×
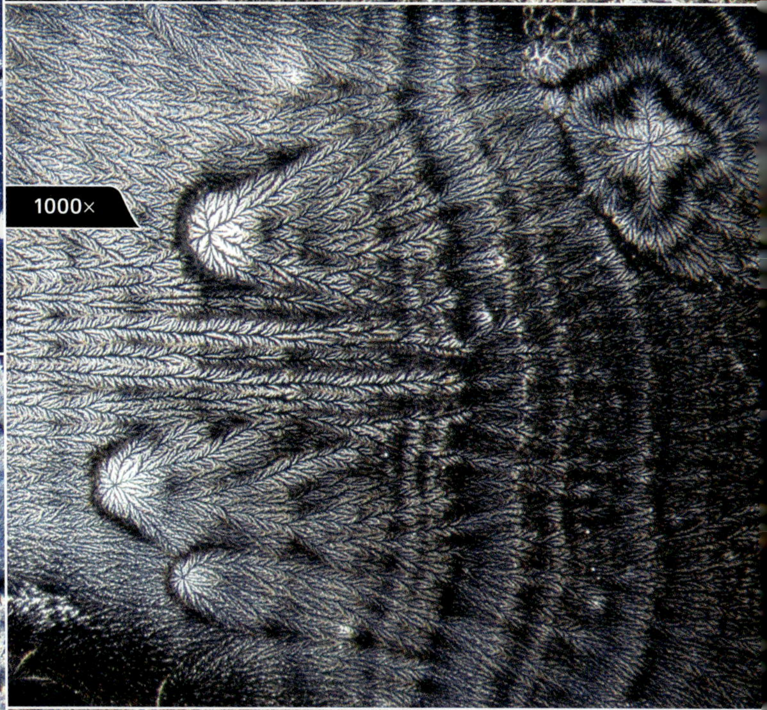
1000×

400×

200×

Canola oil

The order force (life design principle) of the organic canola oil is not entirely convincing. Although in the first example it builds similar formations as the sunflower oil, the "solar cells" of the canola oil are unable to inform the entire droplet and yawning information gaps abound. The information is still there, but it is rather weak. However, in another rather unambitious formation it succeeds in pushing through. That one then is probably closer to the essence of the canola plant. Finally this series also reveals interesting plant-like landscapes with a character of their own.

In the nonorganic canola oil, beautiful plant-like shapes appear only on the borders. Everywhere else the natural information is erased or heavily disrupted. Even when a fragmentary beautiful branch is formed, it is not solidly built and leaves an impression of brittleness; at any moment it may fall apart into loose dots, as has already happened in its surroundings.

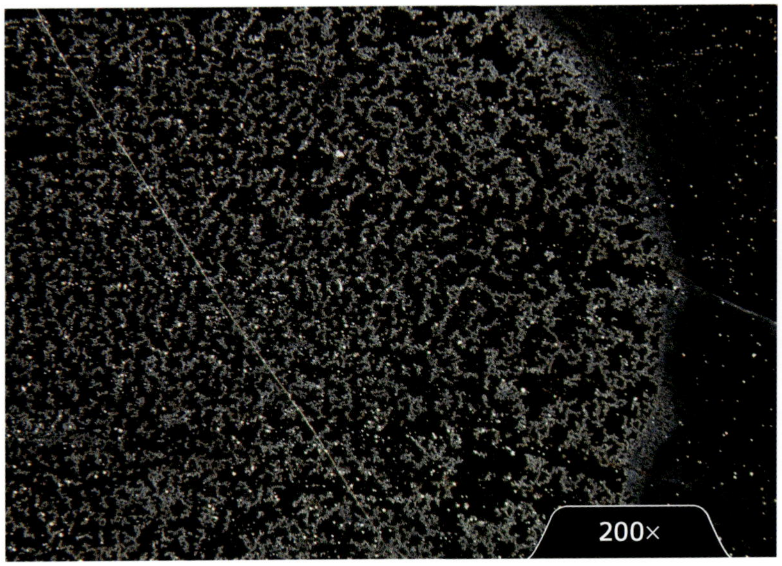

NONORGANIC

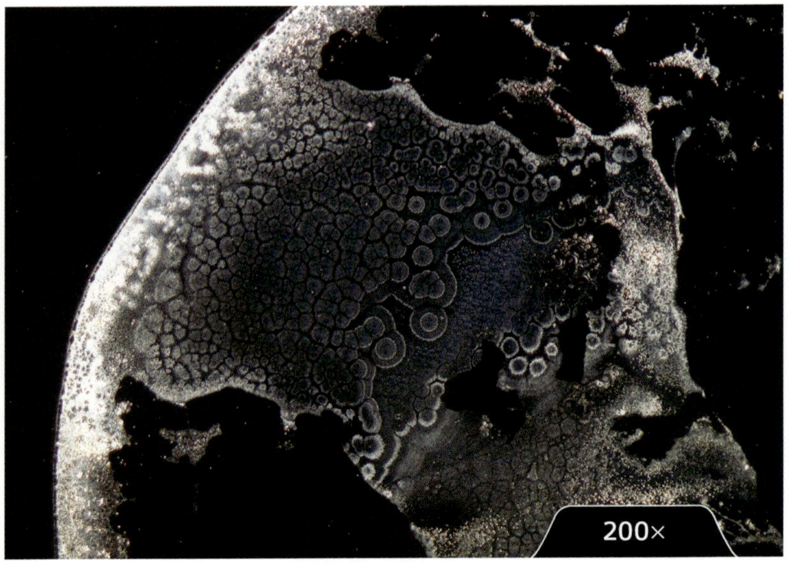

ORGANIC

200×
200×
100×
200×
NONORGANIC
ORGANIC
400×
200×
400×

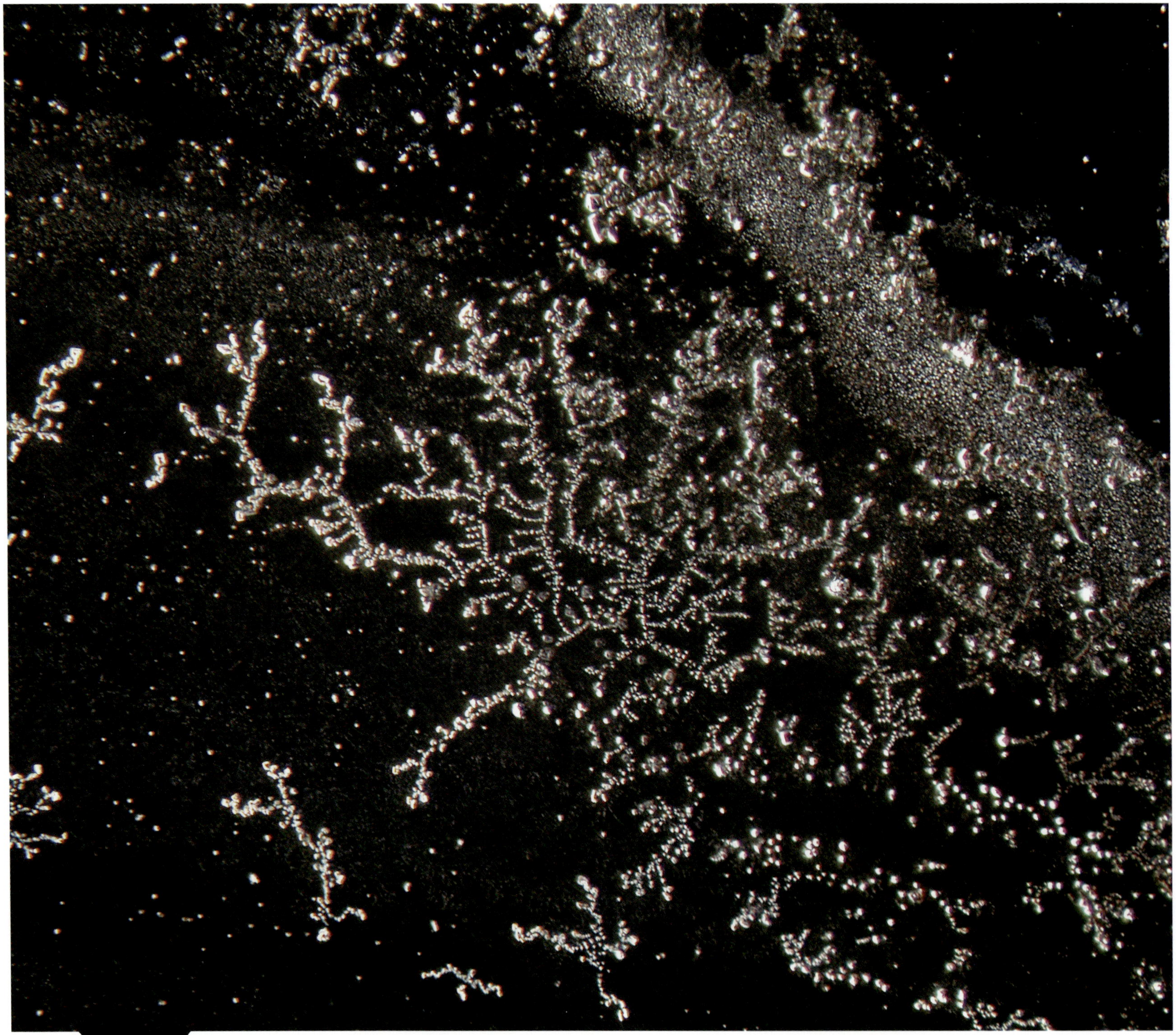

400×

NONORGANIC

ORGANIC

Coconut oil

In the West organic coconut oil has recently become a highly valued food, and for good reason, after having long been dismissed as a fattening saturated fatty acid.

The life-energy of organic coconut oil surprises us with giant flowers and fine wreaths, vigorously and characteristically shaping the droplet, together with a circle of small six-petalled stars.

The life-energy of the nonorganic coconut oil tries to craft the exact same beauty, but it has lost the strength to do so, resulting in information gaps. Instead of harmonious wreaths, a chaotic disorder ensues. It is only on the borders that nicely looking stars can sometimes be found.

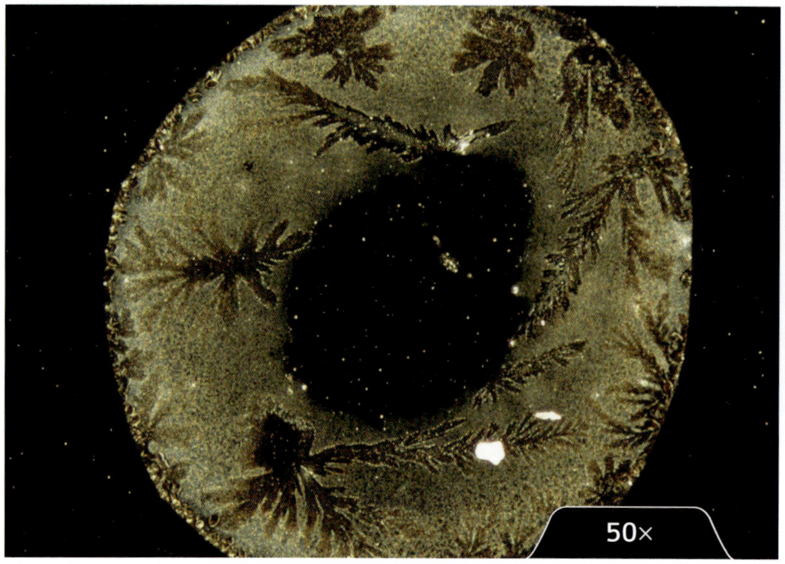

NONORGANIC

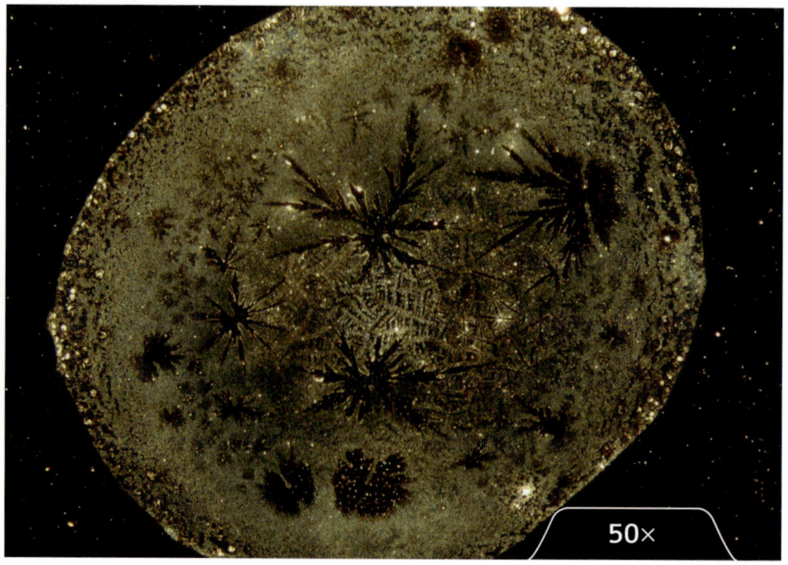

ORGANIC

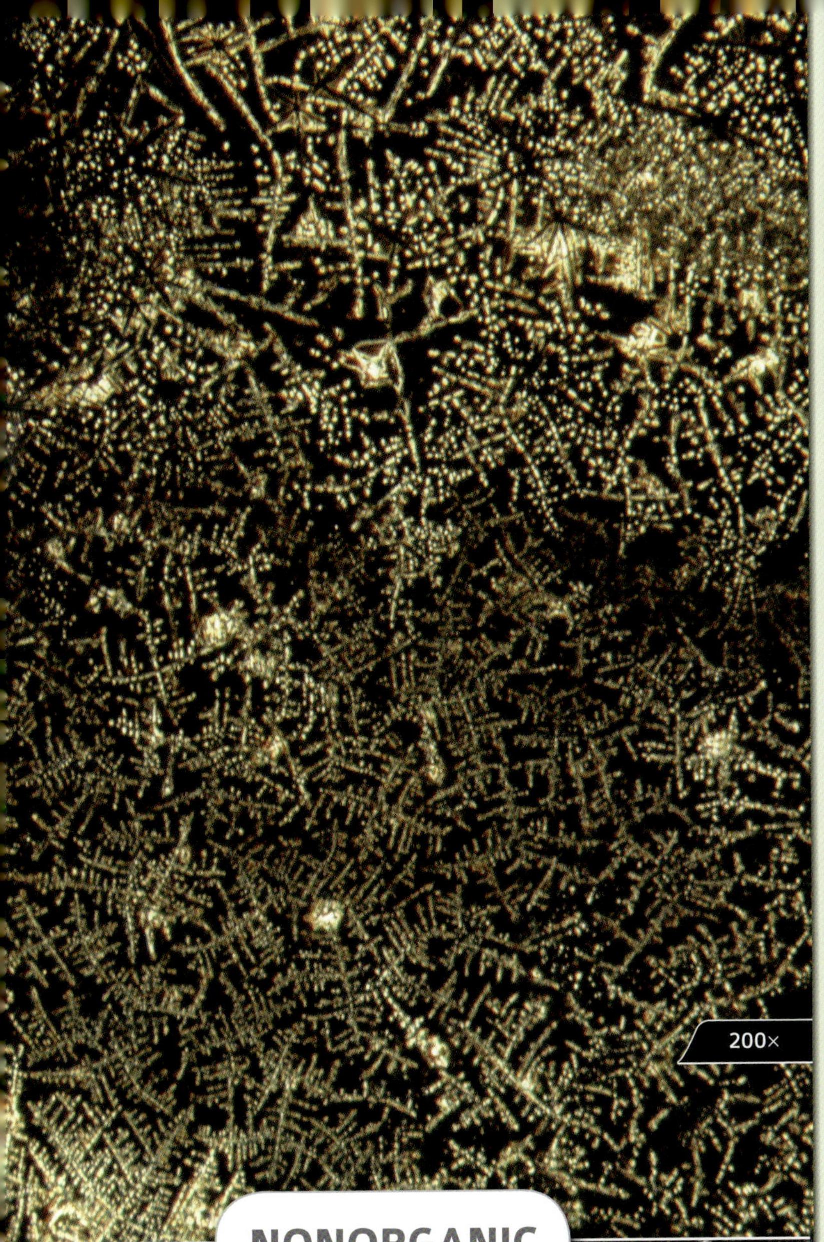

NONORGANIC

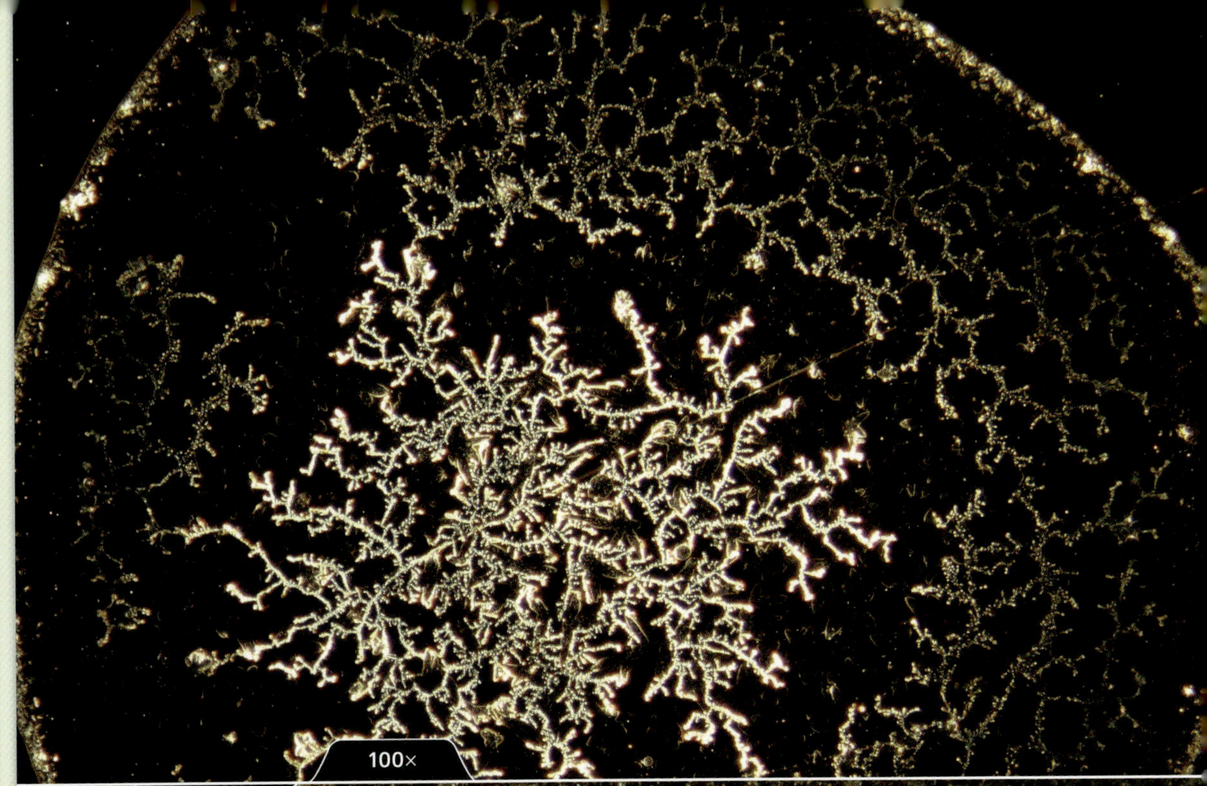

ORGANIC

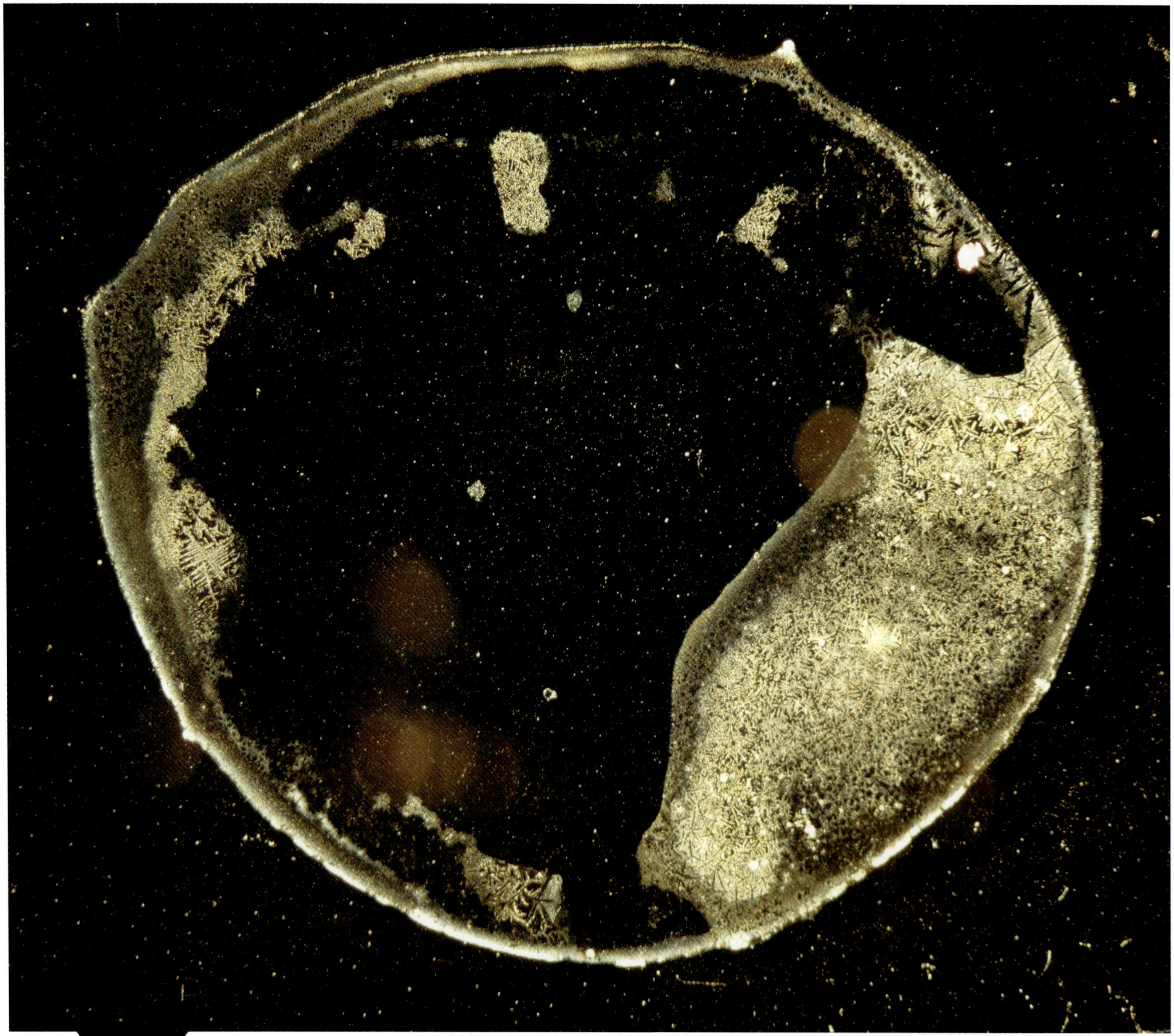

NONORGANIC

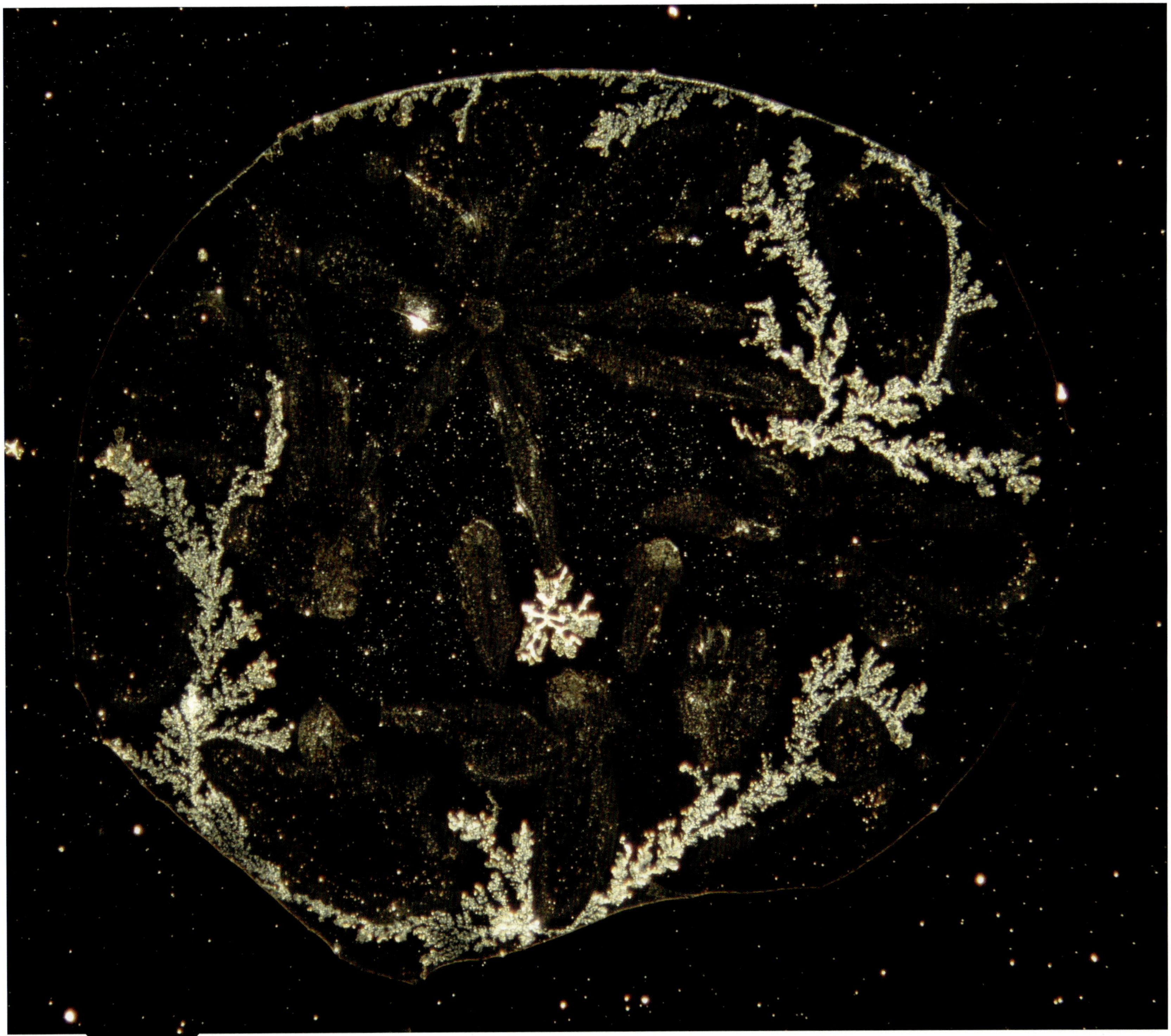

50×

ORGANIC

Alfalfa

The organic alfalfa sprouts have an appearance of powerful rods or supportive struts, but they can also be soft, decorated with vernal blossoms.

In comparison the nonorganic alfalfa sprouts are weaker, with holes and have often forgotten some of their information. Otherwise they are not that different from their organic sisters, possibly because few toxic chemicals were used in their development, unlike in other foods such as grapes.

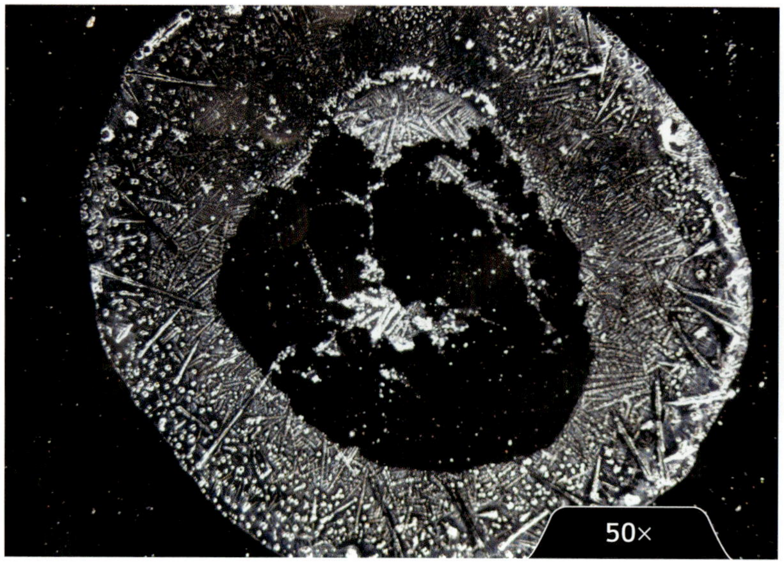

NONORGANIC

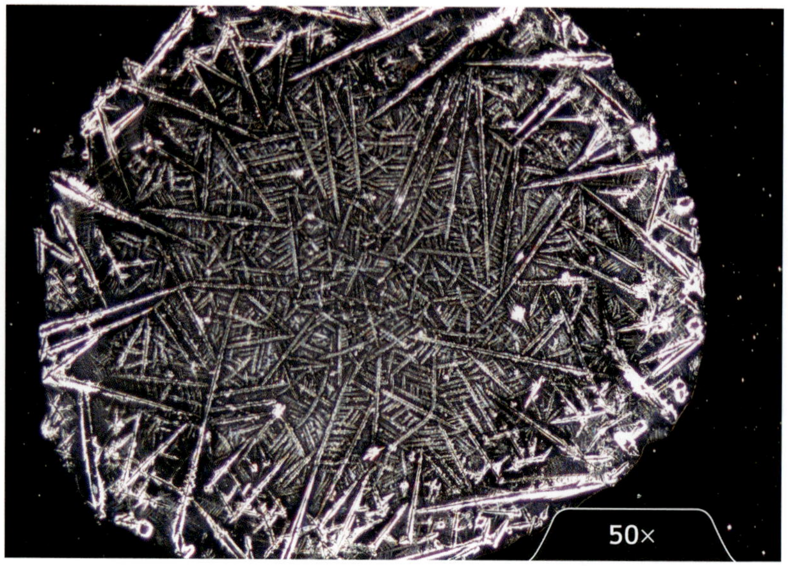

ORGANIC

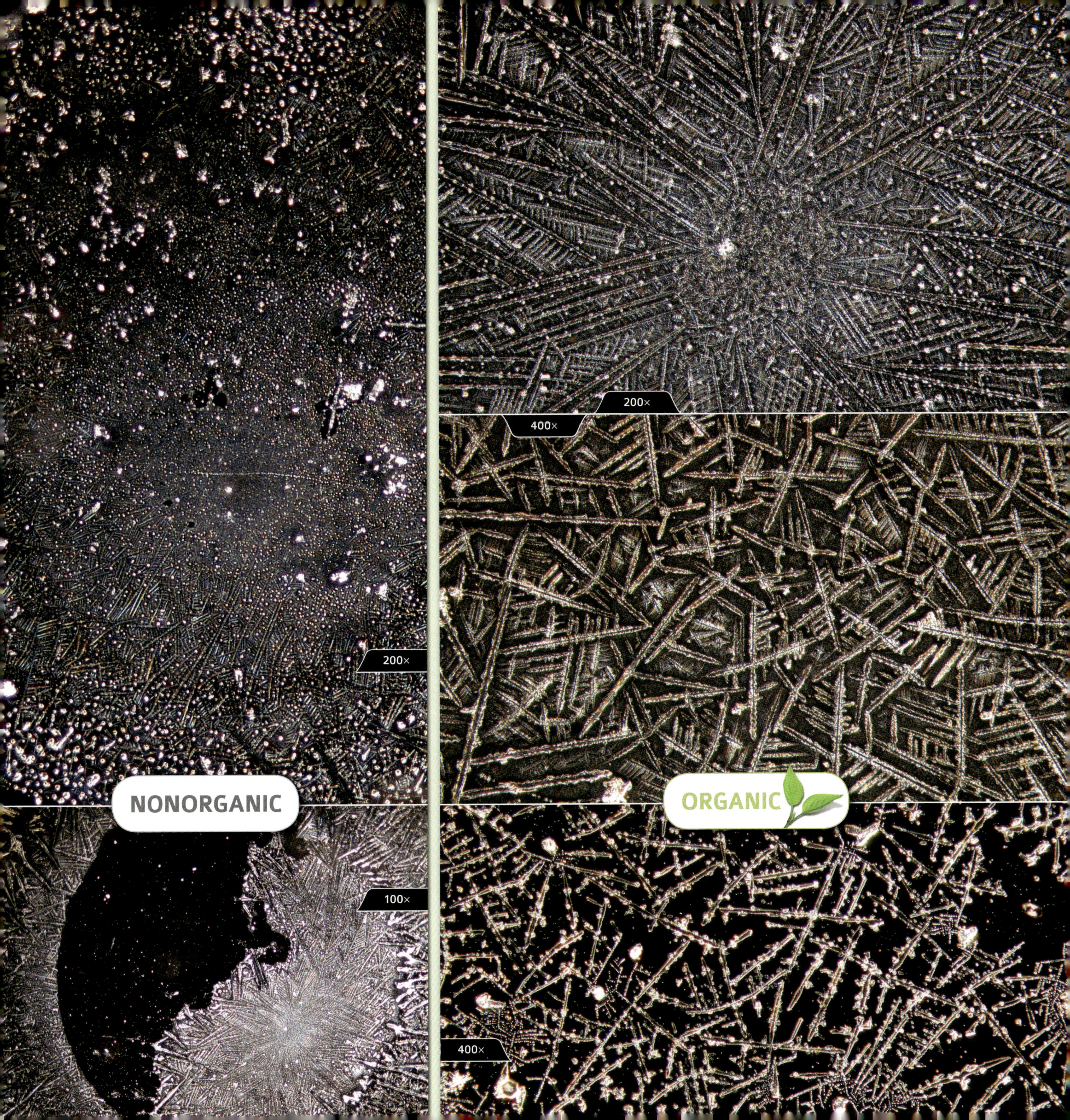

200×
200×
400×
NONORGANIC
ORGANIC
100×
400×

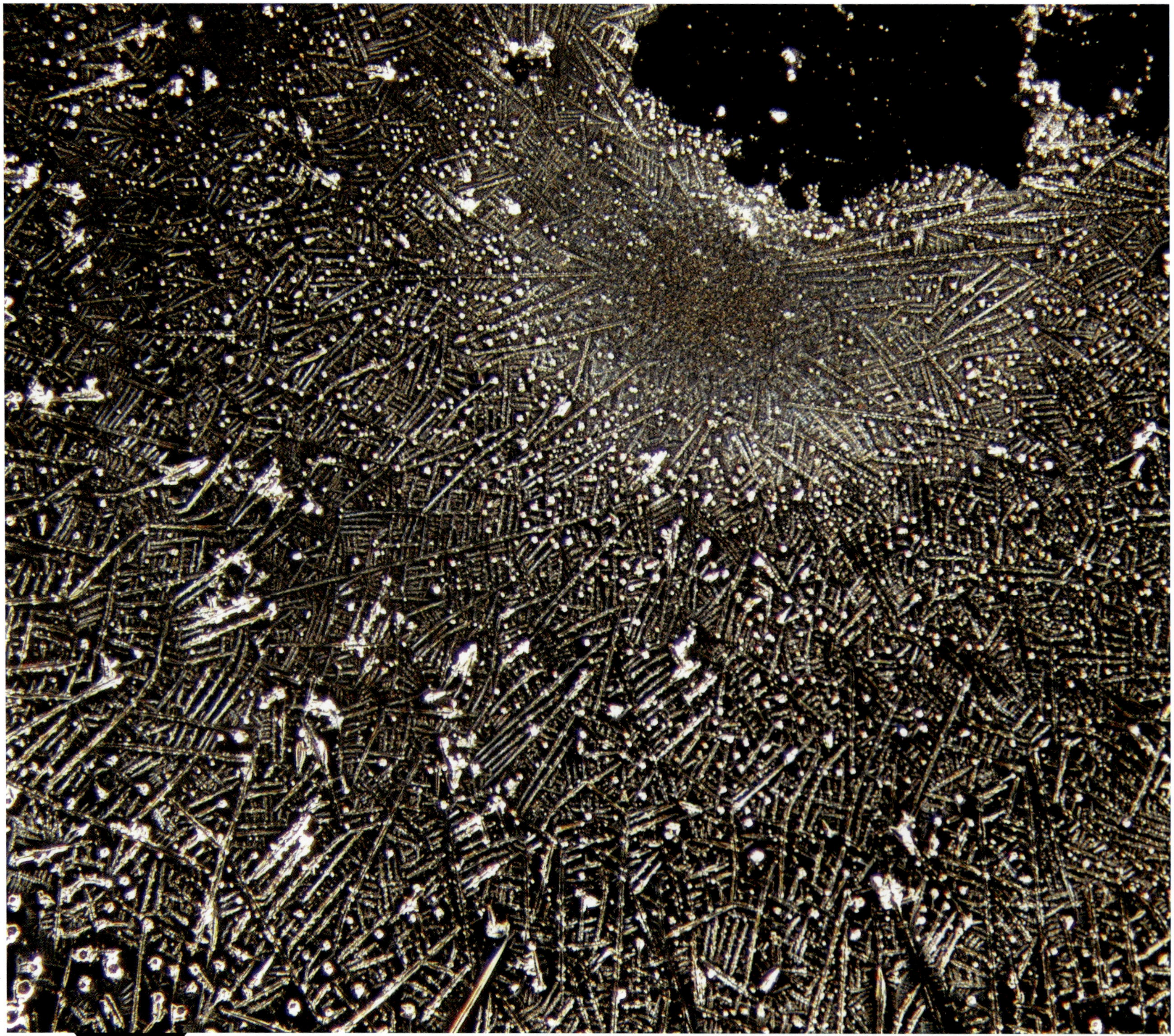

200×

NONORGANIC

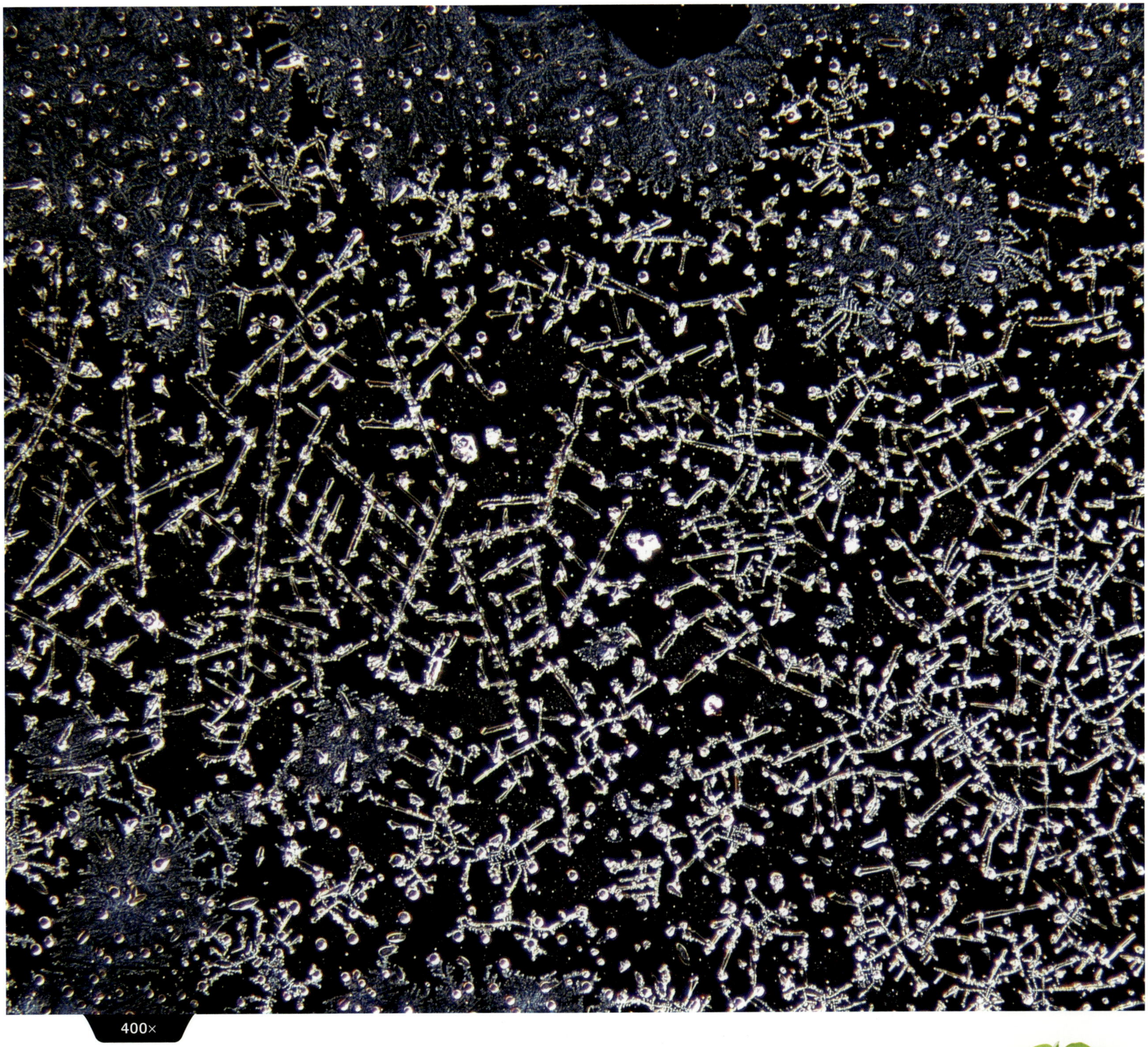
400×

Dandelions from two locations

"Wild dandelions are great and always full of life-energy,"—so we might think. Yet a small test reveals this is not the case. When something grows in the wild, there are many factors on which the development of the plant depends. We summarize these differences here according to the place they are grown, at locations 1 and 2.

What do we see? The wild dandelion from location 1 has a different character from the wild dandelion from location 2. It seems that at location 1 conditions were better than at location 2, because the order force (life design principle) in location 1 builds clear forms and a beautiful centre, whereas in location 2 it looks rather damaged with information gaps and disheveled forms.

What do we learn from this? Wild is not synonymous with good, organic or natural. Dandelion greens picked by the roadside may have been subjected to influences that have stunted its development: polluted air, unclean water, earth containing toxins, or unfavorably situated with regard to the sun.

Therefore, plants grown in the wild are only suitable as quality foods when they have had favorable conditions: clean air, pure water, healthy soil and sufficient sunlight.

LOCATION 2

LOCATION 1

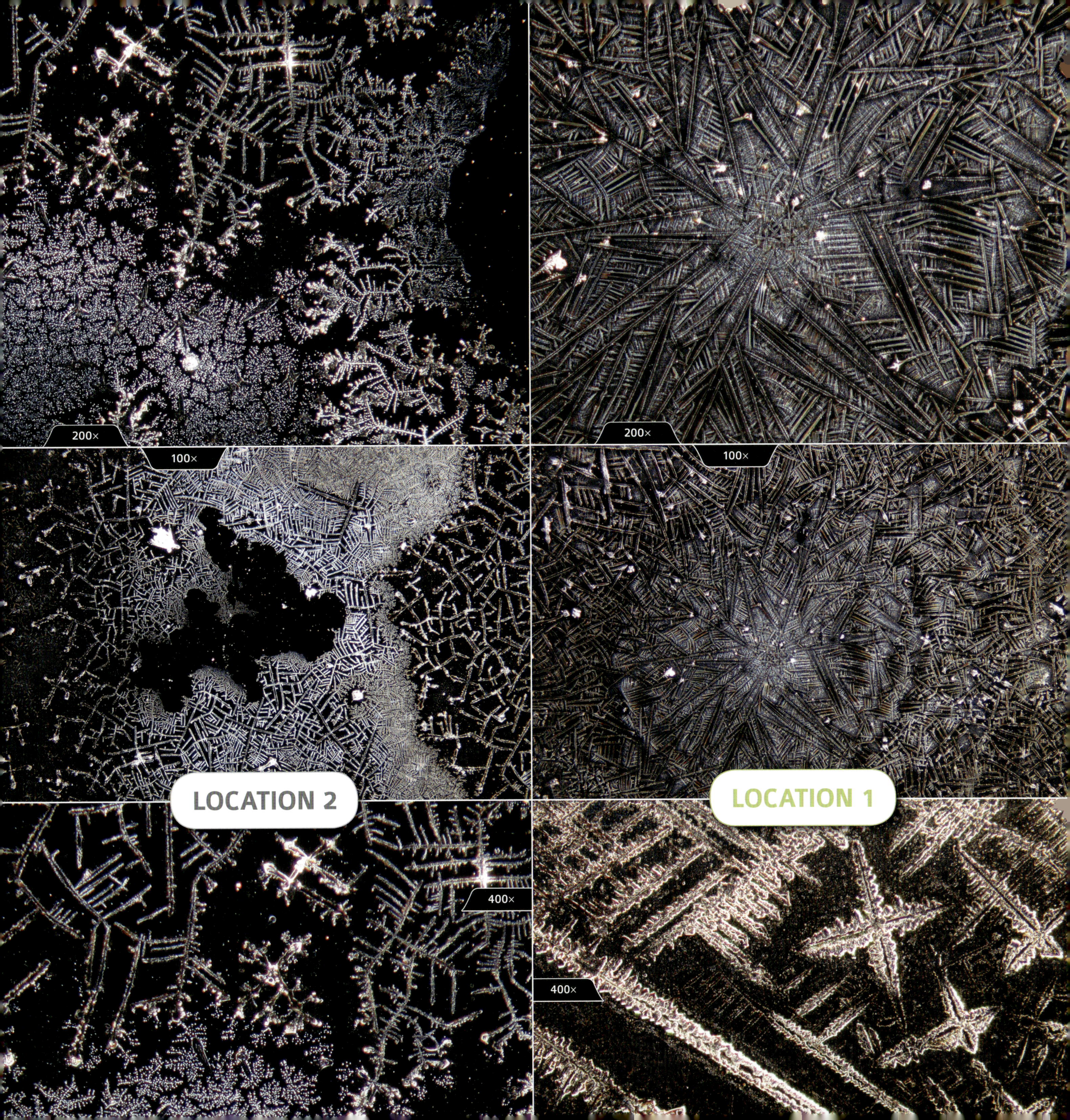
200×
200×
100×
100×
LOCATION 2
LOCATION 1
400×
400×

Wild Rosemary

and the same wild rosemary after 3 minutes in a microwave oven

The order force (life design principle) in wild rosemary forms a solid house with a powerful defense on the outside and very fine lilac bushes on the inside.

The effects of the microwave oven can be compared to a fire: after the house fire, the finely structured inner life is destroyed. The house resembles a charred ruin.

This is what we do to our food when we briefly heat it up in the microwave.

Images depicting the life-energy of food after "warming" them in the microwave have never been seen before. For the first time our eyes are able to see what was previously invisible. Let us ponder for a moment what this means for the millions of people who completely destroy the order force (life design principle) in their food this way on a daily basis, especially for patients in hospitals, where a microwave can be found in every ward, as well as for sick or elderly people living alone who use it to heat up their food. In thousands of restaurants, deep-frozen meals that have been purchased from industrial companies are quickly "regenerated" before serving, in order to save labour. Those working in the health sector, who have the responsibility for organizing, monitoring and guaranteeing food safety and quality, may wish to ask themselves whether this method is really so harmless that it can be used unreservedly and furthermore, whether microwaved food should be offered to the consumer without any disclosure. Or does the attitude-change have to come from conscious consumers, who should inquire as to how their meal was heated? There seems to be room for welcome improvements here.

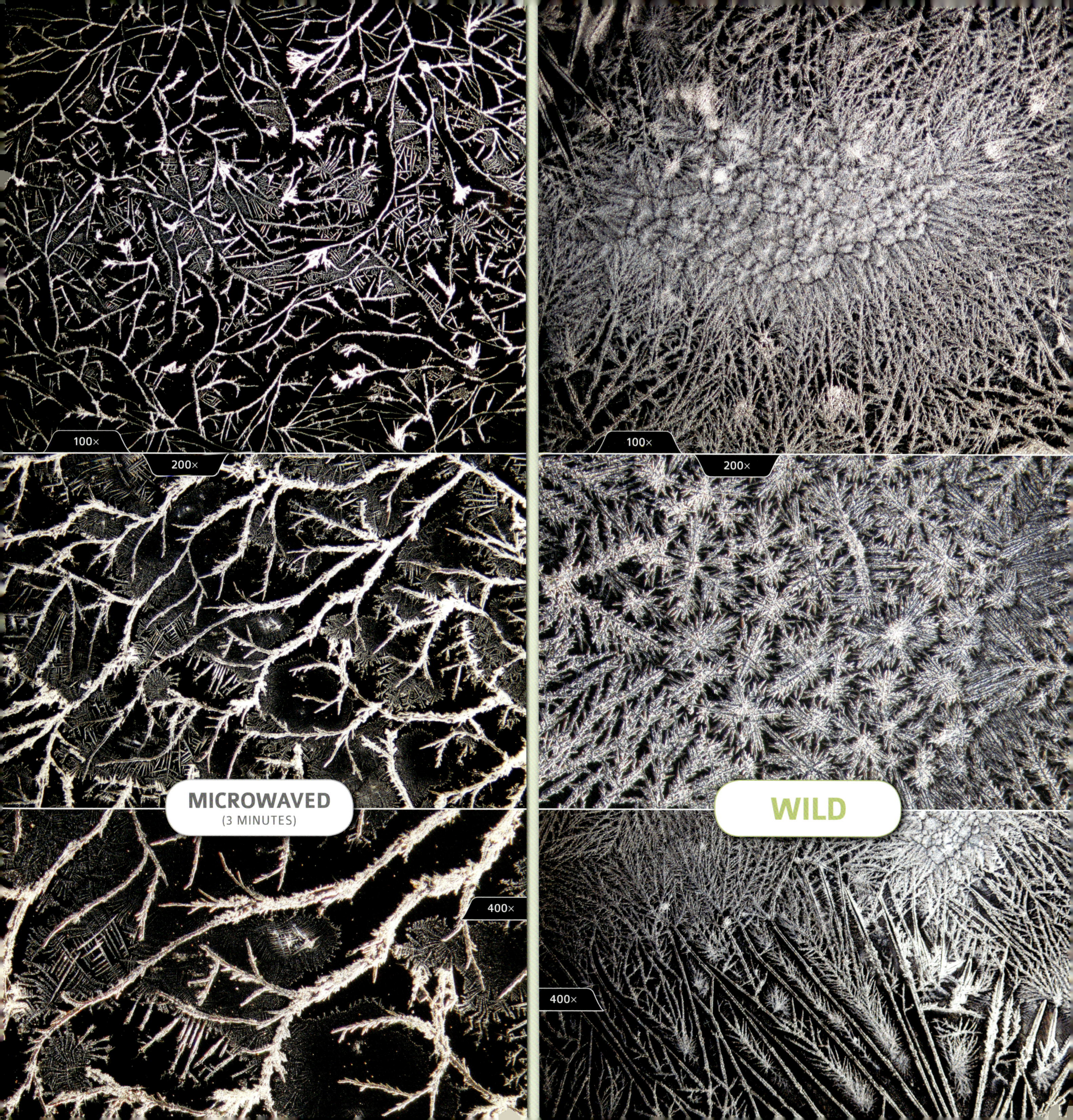
100×
200×
MICROWAVED
(3 MINUTES)
400×
100×
200×
WILD
400×

400×

MICROWAVED
(3 MINUTES)

MICROWAVES

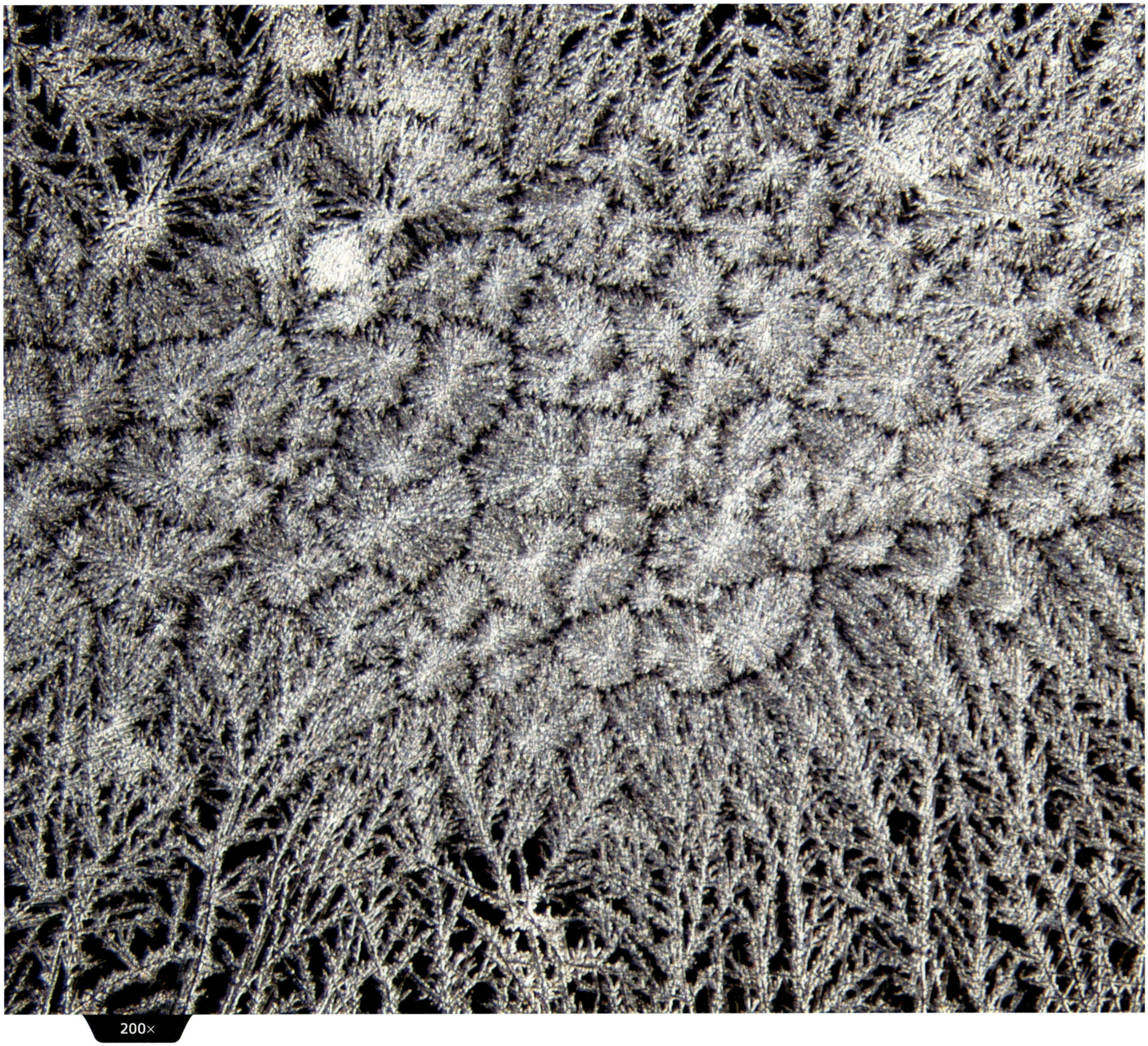

WILD

Soyana foods

Since we founded the LifevisionLab in 1996 for internal quality research, it naturally follows that we first applied the standardized Soyana Method in this new laboratory to our own foods. In this way we were able to get a better understanding of the invisible qualities of our raw materials and manufactured foods, to investigate the influences of storage and processing on the organic ingredients, and to verify our efforts for quality improvement.

Our goal was, is and will remain the production of organic foods with the highest possible quality for our customers. The fact that the members of the Soyana team daily pray, meditate, practice sports and also regularly consume our own manufactured foods, has resulted in a dedicated search in all directions as to how we can produce the best and finest foods. True to the motto in our name "The Secrets of Perfection-Flames", the search for improvement is incorporated into the very fabric of our company and forms an integral part of Soyana's corporate culture. That does not mean we never make mistakes, but we consider mistakes as opportunities that reveal previously unnoticed aspects of quality improvement. For our team this has led to a constant process of learning and improving for its own sake, which ideally turns into a spontaneous path to self-realization through insight and interest.

Although in this small section of the book no comparisons are made between organic and nonorganic foods, it may still interest the reader to see, by these few examples, what kinds of invisible forces are at work in Soyana foods, enabling us to offer a small contribution to the wellbeing of our customers.

1		2	
3		4	
5	6	7	8

Page right:

1 Chi – Organic Fermented Drink: Flower-Chi // **2** Organic Tofu Burger: Vegetables // **3** Swiss Cereal Drink: Organic Millet // **4** Swiss Rice-Drink: Organic Brown Rice Plus Calcium // **5** Chi – Organic Fermented Drink: Herbs-Green Tea-Chi **6** Chi – Organic Fermented Drink: Root-Chi // **7** Chi – Organic Fermented Drink: Fruit-Chi // **8** Organic Tofu Burger: Fit (Sprouts)

50×
200×
200×
500×
ORGANIC
200×
1000×
400×
200×

200×

ORGANIC

Soyana Organic Tofu: Natural

200×

ORGANIC

Organic Tofu Burger: Vegetables

Soyana Dinki (meat alternative from organic spelt)

400×

ORGANIC

Soyananda Vegan Soy
Yoghurt: Natural

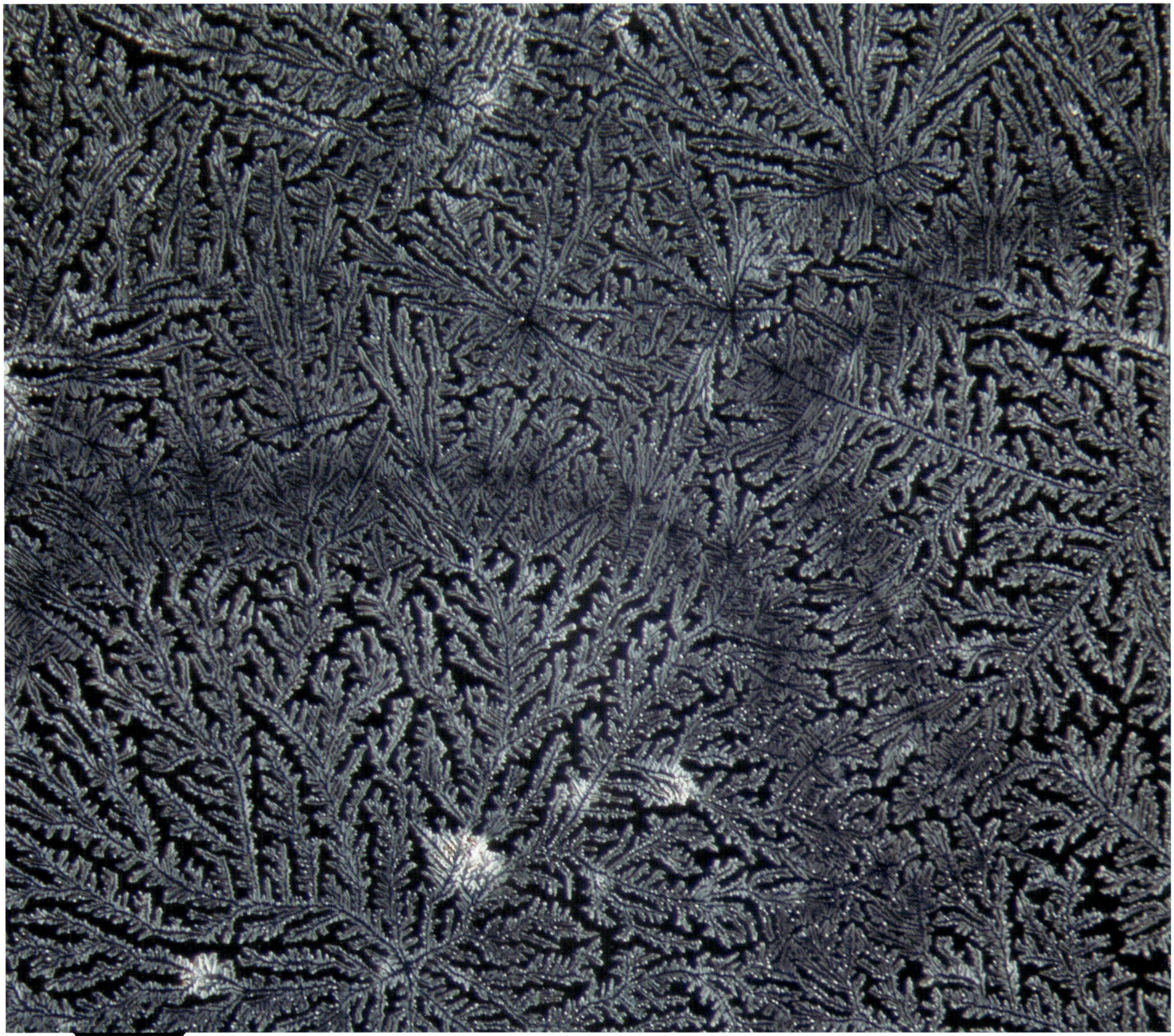

200×

ORGANIC

Swiss Cereal Drink:
Organic Spelt

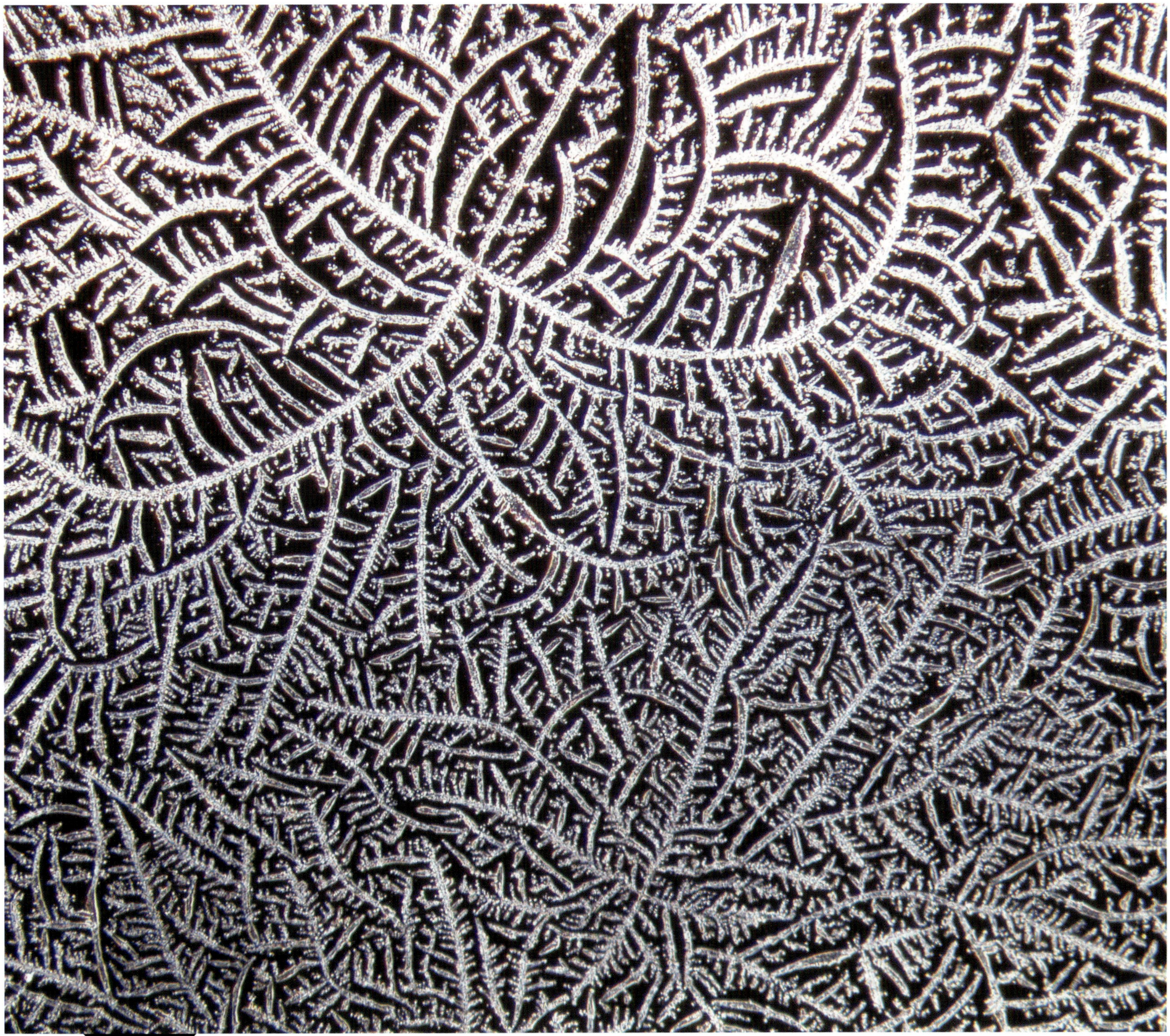

400×

ORGANIC

Swiss Cereal Drink:
Organic Millet

1500×

ORGANIC

Chi – Organic Fermented Drink:
Fruit-Chi

400×

ORGANIC

Chi – Organic Fermented Drink:
Root-Chi

The obvious message in foods

What do the images of the order force (life design principle) in foods tell us? The questions we asked at the beginning of this book can now be answered. We do not need 200 pages to observe that organic and nonorganic foods differ significantly in their life-energy or order force, and that these observations warrant the following conclusion with a huge correlation:
In the expression of their order force, organic foods are characterized by a greater degree of clarity, beauty and order. Nonorganic foods are weakened in their order force.

Before publication we conducted a small survey using the photographs in this book. The following are the observations of the vast majority of the participants:

Nonorganic foods, with regards to their order force, are generally more...

- ugly
- weak
- coarse
- unharmonious
- artistically unfinished
- blocked
- aesthetically unconvincing
- unnatural
- hostile to life
- incomplete, fragmentary, "forgetful"

... and therefore offer a reduced value

Organic foods, with regards to their order force, are generally more...

- beautiful
- strong
- finely structured
- harmonious
- artistically complete
- flowing
- aesthetically convincing
- natural
- friendly to life
- complete, connected to the whole

... and therefore offer a wholesome value

Even if these details do not apply to all points in every example, we are able to see at least some of these differences very clearly most of the time.

The influence of chemical-industrial agriculture on the natural order force (life design principle) of foods is definitely severe, as we have often seen the beautiful organic forms being extinguished in nonorganic foods to a greater or lesser degree, or pushed to the sides, often leaving information gaps or a kind of sedimented waste in the centre that is unable to express the original information. This blurring or disappearance of the natural memory is reminiscent of the symptoms of Alzheimer's disease, where the brain of the patient no longer stores any memories and is left with a yawning emptiness.

» *Organic foods are full of beauty and power in their order force, and therefore they are valuable for our nutrition.*

The order force of organic foods proves to us that organic foods are able to deliver nature's original information, beauty and power to the consumer without fail, confirming their ideal and health-promoting role in the nutrition of humans and animals.

» *Nonorganic foods are weakened in their order force, and therefore their nutritional value is reduced.*

The order force of nonorganic foods shows us that nonorganic foods have lost nature's original information, beauty and power to a large extent. In the long run they are unsuited as health-sustaining nutrition for humans and animals, and should perhaps only be used in emergency situations.

According to food safety regulations of the official food agencies in Switzerland—and probably in Europe and the US as well—a food product that is "reduced in its value, is no longer marketable", as the terminology states, meaning they should no longer be offered for sale. Due to new laws this includes food for both humans and animals.

Naturally it will take some time before the order force (life design principle), invisible to the naked eye, will be officially recognized as a quality criterion, through pressure from further research findings and consumers. Yet it is clear that as soon as the criterion of the order force in foods is recognized, today's high standard of food safety will favour organic foods.

The correction of a grave error

The methods of ensuring food safety, and judging food quality and the effects of farming on ecology and health, require correction, which would be the correction of a grave error.

In fact it is time for a complete change of perspective in this area. Until today the adoption of chemical-industrial agriculture has been scientifically justified on the basis of conclusions from the research findings of Justus von Liebig during 1839-1849, on superphosphate and other chemical additives. It was he who discovered the effects of dissolved minerals on plants.

The following article (Wikipedia in German, accessed 7 August 2016 and translated into English) provides the following interesting history and describes the motivation for his discovery:

> His (Justus von Liebig) main interest during his time in Giessen was the promotion of agriculture with the aim of preventing the sometimes devastating famines of that time - in 1816 he had experienced one himself in a year without summer. He published his knowledge in this field in 1840 and 1842 in two works: The Organic Chemistry in its Application to Agriculture and Physiology, shortly named "Agriculturchemie", and The animal chemistry or organic chemistry applied to physiology and pathology. These two books aroused enormous attention, not only among scientists, but among all educated people of his time. The agricultural chemistry, in which he advocated the use of mineral fertilizers and explained its importance for the quality and yield of plants, was published successfully for nine editions and was also translated into 34 languages.
>
> In his private laboratory, he devoted himself from 1846 to 1849 to the development of a water-soluble phosphate fertilizer: the result was the so-called super phosphate which today is still the world's most widely used phosphate fertilizer. The fertilizer greatly improved harvests and thus the food supply in the second half of the 19th century.

Today we can appreciate Justus von Liebig's achievement as a welcome solution to short-term emergencies like famines, which was indeed his prime motivation.

But unfortunately chemical-industrial (conventional) agriculture does not result in holistic, wholesome foods for the middle- and long-term nutrition of people in normal times. For such purposes the order force (life design principle) of these foods falls short, being clearly damaged by the use of artificial fertilizers and chemical additives.

From the current holistic point of view, the scientific opinion (which originated in the 19th century) that artificial fertilizers and chemical additives are indeed a valid method for cultivating wholesome foods can be considered as one of the gravest scientific errors of modern time. Its implications have been immeasurable.

In the 20th century, chemical-industrial crop cultivation with its huge increases in productivity in the agricultural sector became a globally accepted and practiced convention, hence the name "conventional agriculture". In the late sixties during the so-called Green Revolution the method was transferred to developing countries on an enormous scale.

Yet today the methods of ensuring food safety and judging food quality must be expanded and their earlier assumptions investigated. Based on the results of the research documented here, nonorganic foods are not the wholesome food products they are mistakenly thought to be, based on their past scientific assessment. Needless to say further research is needed.

Justus von Liebig (1803–1873)

The shock of experiencing famines even in the 20th century led to the understandably strong reaction of chemical-industrial agriculture: a battle against nature, which can never be won in the long run, since nature is infinitely stronger than man. Organic agriculture is the correction of the overreaction through a sensible cooperation with nature. Man no longer places himself above nature, but recognizes himself as part of nature.

The importance of creating awareness

Creating awareness is an important or even crucial part of careful preparation for applying the insights we have acquired. By revealing the hitherto invisible power of foods, we have obtained new knowledge, which sets the stage for a big transformative process in the food industry, including its science and education.

When the results from the Soyana Method are confirmed by further research, which we deem very likely as we have achieved these results through clean and reproducible methods, a new and holistic definition of food quality will present itself: one that includes the order force (life design principle).

We need courage in order to set new goals that correspond to our level of understanding, and we require wisdom in order to win people over to embrace these goals for improved wellbeing and to joyously achieve these goals for everyone.

It is therefore necessary to discuss a new evaluation of the nutritional situation among the public, and also to discuss the farming methods that are generally practiced and taught in public schools.

Based on the results documented here and based also on the human right to healthy food, it is in the best interest of consumers that in the future, all humans and animals should normally be nourished with organic foods, i.e. foods that are wholesome in their order force (life design principle). Their order force should be natural and not reduced in its value.

The human right to healthy foods means ORGANIC FOR ALL.

In order to achieve this objective gradually and harmoniously, an approach is needed that is built on insight and a deep, natural understanding by both producers and consumers.

Since we all prefer to improve our actions through insight rather than under the compulsion of sanctions and regulations, it is useful to ask: what is the best way to cultivate and promote insight among all participants?

It is important to prepare carefully for the big changes ahead, through the process of developing awareness, and to implement these changes with inspiration, gradually and voluntarily.

Health education and promotion by means of excellent illustrative examples, which inspire positive and strong emotions, are important. Inspirational advertising campaigns, created by the best advertising agencies in the country and seeking to raise awareness, are far superior to new legislation. These should be the means for implementing modern economic and health policies.

Is it possible to become aware of the kind of thinking, the mentalities and the unquestioned assumptions that have shaped our current behavior with regard to how we nourish ourselves?

A good example of organic advertising that includes positive emotions and fun: TV campaign for Coop Naturaplan (2012) in Switzerland. Link to the video: *www.lifevisionlab.org* under *Links*

Which underlying models have shaped conventional food production? Is it possible to recognize these models of thought, which affect our behavior so strongly, and improve them through deeper and better understanding?

Are there any significant elements in this field that harm us and from which we can free ourselves? Are there any significant elements that can promote our wellbeing?

To probe more deeply into the subject, we ask ourselves the question: what *is* food, in reality?

What is food, in reality?

Our programming is outdated.

The old model says: food is fuel for our engines. Calories seem to be very important! That is why foods are identified first and foremost by their calorie content (cal./Joule). This is the energy released when a food is burned. Wikipedia gives the following definitions of 'nutritional value' and 'energy carriers' (translated from the German edition, September 18th 2015):

The nutritional value is the main benefit of food. It is a measure to qualify and quantify the physiological energy or caloric value of food. The term 'nutritional value' usually only applies to the energy value - the calories provided to the body.

Energy carriers: nutritional substances supplying energy to the body, which when successfully converted provide also building blocks for physical growth and regeneration. These essential nutrients are composed of proteins, fats and carbohydrates. These components of an individual foodstuff primarily provide the body with energy. Therefore they are also described as fuels.

So, are foods a kind of fuel for our engines?

Yes, but only at the very end, after our organism has used the food in many other ways and has no further use for it.

And what happens first when we put something in our mouth?

First of all, the subtle energy of the food enters into our subtle body. The food first gives us its life, its wealth of information, its capacities, its knowledge, its order force (life design principle), its memories and experiences. All these are stored in its subtle body. Now we understand something new:

» Foods are highly developed information systems that sustain life.

Is our mechanical model of food as mere fuel for the engine not a little one-sided and outdated?

Indeed: the mechanistic model of 'engine fuel' is outdated.

Imagine quantifying the value of a computer by its energy value in calories, which we would measure by burning the computer. This metaphor illustrates how outdated the mechanical model of food is.

Everybody knows that a computer is an information processing system. But most of us still think of foods in terms of fuel. As the computer is a device to process information, natural foods are systems that are infinitely more intelligent, transferring information and even consciousness to us in order to sustain our life, long before they provide us ultimately with their fuel energy.

Yet we still take our cues from the outdated mechanical model. Indeed, we only see a small part of the whole reality, the material part, and miss the larger part. And we do not even notice it. To put it bluntly, **we nourish ourselves as if we were (gourmet) engines!**

The consequences are manifold and quite tangible: the subtle or information-containing part of food remains unnoticed, weakened by chemical-industrial agriculture and the modern world: its life, its wealth of information, its capacities, its knowledge, its order force (life design principle), its memories and experiences remain uncultivated and increasingly fade away. Even worse, they are being replaced by negative content. The inner quality in food is a neglected subject today, whereas its value as fuel and enjoyment attract attention.

We are producing food according to an outdated mechanistic-materialistic model.

» *The mistakes created by the outdated model may prove fatal.*

The currently prevailing understanding of foods and their nutritional value is the result of a one-sided materialistic mentality and its mechanistic model. As documented in this book, this causes the chemically and industrially produced foods to...

- contain less life-supporting information,
- show increasing weakness and chaos in their subtle or information-containing components,
- decline in their order force (life design principle), and
- become increasingly negative in their order force.

Those who nourish themselves with these foods receive...

- ever-decreasing support from the subtle, information-containing component of nutrition, which should promote good health, and
- ever-increasing hostile influences which erase the natural knowledge—the biological memory and capacities—on the cellular level.

It is possible and even likely that, for these reasons and other influences arising from our lifestyles, our disease susceptibility is increasing, posing a growing threat to our health and viability.

Needless to say, statements on health effects must be backed by further studies, but it is clear that allowing unchecked hostile influences to affect our sources of nourishment is a dangerous game, which we are now playing in a big way—even though this is mostly out of ignorance or lack of interest. The very idea of food safety does not permit such an approach.

The phenomena observed in the current studies—that conventional agriculture drastically decreases the order force (life design principle), the information in the cells and the biological memory of foods—are alarming and sound a note of caution when dealing with so-called conventional methods of food production. In reality these methods are grave intrusions into the vitality of our foods.

Once we become mindful of the need to preserve the health of humans and animals, then the time will come to change the model to a new one:

The realisation that we are much more than a mere material mechanism.

The new model is the result of a revealing vision of the whole reality:

» *Each human being comprises a whole and living reality, made up of a physical and a subtle body, a mind, a psychic organ also called the spiritual heart, and an essence also known as the soul.*

All these parts of our reality need nourishment. When produced and prepared with love, food can nourish our whole being. Foods are able to absorb the rich influences of nature through earth, weather, landscape, sun, moon and stars, as well as human emotions, words, music and even consciousness, and transmit these to the consumer. Foods transmit everything that they experience during their life journey to the people consuming them.

The following positive examples show that foods originate from a living relationship with their environment and with the human being:

- Garden plants that experience love and concern from us, thank us through the development of special substances that we need.
- The love that a mother or father feels for their children while cooking makes the food they prepare exceptionally good.
- My aunt from Adelboden in the Swiss mountain region of Berner Oberland used to play music for her 12 cows. She told me the cows gave much better milk and were much healthier than when she did not play music.
- My spiritual teacher Sri Chinmoy told me that for a cook it is most important to prepare the dishes in a high consciousness.
- The examples shown in this book of Mr. Mercandelli's wine and Soyana's fermented Chi drink demonstrate that loving care can create foods that go far beyond the beauty and power of their organic ingredients.

The new model offers a perspective on perfect food quality:

Since perfect health means infinite wellbeing on all levels of reality, therefore the most wholesome foods are those that can nourish us on all levels of reality with their endlessly fulfilling wealth of experience.

Simply put: the life in our body is not visible on the material plane. Nonetheless the life inside us is a very important part of our reality. The life inside us also needs subtle nourishment: life-energy, inner light, consciousness.

Humans, farm animals and pets need naturally grown foods that are both materially and subtly wholesome and are able to support our life.

There are foods that fulfill this need on a fundamental level: organic foods.

The role of organic agriculture and horticulture

Organic agriculture and horticulture are the appropriate food production methods for the nourishment of humans and animals, according to quality evaluation that is based on food safety, ecology and health, and that has the objective of bringing about complete wellbeing.

Today many people and families have already arrived at this understanding themselves. The results presented here further support this understanding, and strengthen the existing tendency among a growing number of consumers to nourish themselves increasingly with organic foods. This development is happening on a completely voluntary basis, through normal market mechanisms, and should continue in this way.

Since many public institutions have come to a similar understanding, with some already establishing programs such as school catering that is based on organic food, society will follow with great interest the professional debate about the new implications arising from the results presented here.

Consumers should decide for themselves

Naturally, consumers can decide for themselves about the quality of the things they buy, but modern economic and health policies should inform consumers in a positive and inspiring way about their options and what effects their choices have, both on their health and on the world.

Through the results presented here, many people will reach a new understanding about food; they include food and health experts and consultants working in academia or in industry, law and policy makers, public officials, members of consumer organizations, as well as food producers and food merchants.

It is extremely important that managers and policy makers pursue a modern economic policy utilizing convincingly presented information, good advertising campaigns, psychologically and educationally mature training methods and awards, so that appropriate signals can be received and acted upon by the public, and followed willingly.

Since this book can only open up a window on new insights, further research is needed to develop the proper information materials to support the agenda outlined above. I am willing to develop research programs with interested parties and to certify their methods.

"The Organic Place" of the Whitmuir Farm (Scotland)

How consumers can make informed decisions

Consumers can be empowered to make informed buying decisions, by letting them see the life-energy or life design principle of the foods they are buying.

On the basis of a brief investigation, I conclude that a majority of consumers would welcome such an opportunity and would gladly make use of it.

Here is an easy way to make this happen: innovative suppliers of organic foods can voluntarily attach a code to their labels or packaging material. Using their smart phones to scan the codes as they shop consumers can immediately retrieve an image of any particular food, prepared by the Soyana Method. They can also retrieve images later, on their computers at home.

To support this process, an independent online platform hosting LifevisionLab images can be created. The platform can be used by food producers and suppliers interested in applying the Soyana Method to examine their foods and willing to share the results with the public.

(Parties interested in the examination of food can contact us at info@LifevisionLab.org, or by calling +41-44-731 12 26, LifevisionLab, Turmstr. 6, CH-8952 Schlieren, Switzerland.)

Upon agreement with the food producer, images of all the foods that have been examined can be stored on this independent platform, where customers can search for them. In order to support this objective, Soyana has made the website *www.LifevisionLab.org* available, where LifevisionLab images organized by producer can easily be found. Innovative food producers can now inform their customers in an objective way about the quality of their foods. It is our intention that authentic, reliable, and high-quality information will thus become publicly available.

Is it really organic? Protection from fraud

In 2008 and 2009 the entire organic scene in Europe was shaken up by a fraud scandal, in which a criminal gang, operating under the cover of a new Italian organic trading firm, was buying up nonorganic foodstuffs to resell them all over Europe labelled as organic foods. The criminals were attracted by the price differences between nonorganic and organic foods, until the fraud was discovered by backtracing controls. I had an interesting conversation on this topic with a dedicated lady from BioSuisse, the federation of Swiss organic farmers, who confided in me: "You know, the entire organic control system is based on the assumption that all participants fill out the control documents truthfully, and that the organizations for organic quality control then of course verify with the producers and processors whether the data are correct. Eventually everything will be detected, but there is no absolute certainty in the moment as to whether something is really authentic. The most important foundation is the trust between the people and companies that have built personal relationships over many years."

I answered her: "Yes, among the pioneers that is one hundred per cent true, because quality control checks always confirm their real attitudes again and again. But over time new people and companies join, some of whom are not supporters of organic foods out of conviction. In the fraud affair it was this very issue that shocked everyone, because the criminals systematically falsified everything and additionally smeared some people.

"However, we think there are two tests that offer pretty good protection against fraud, as the results of these tests will fall within a well-known area if the foods are organic: a residue analysis of pesticides in a qualified analytical lab and an examination by the LifevisionLab of the life order force within the foods. If the raw material in question, such as wheat or soy beans or orange juice concentrate, etc. had been examined earlier, we would have had empirical values that the newly examined food should correspond to. When the results from both tests are within the known range for organic foods, we can rest assured that the food is most likely organic. And when the pesticide residue value is markedly higher and the images show a life force that is diminished and disrupted, then we can also know with considerable certainty that it is a conventional food.

"Therefore, we are not as helpless anymore as we were in the past, and we can provide a realistic assessment within a few days."

I also told her about my book project and she was very interested to learn more, because it provides new, additional methods for cross-checking. Today we feel fortunate to be able to share our experience of more than 15 years through thousands of examinations in the LifevisionLab and to offer some certainty to help resolve doubts about whether a food product is really organic, or to help advise when, for example, a food manufacturer would be better off not buying a shipment of raw materials from a new contact. We were actually able to tell a processor of organic rice that, based on an examination in the LifevisionLab, something seemed to be wrong with his organic rice. The follow-up inspection promptly discovered a mistake with one of his suppliers.

Since 1990 the Soyana Team has been working in this historical building in Zürich-Schlieren, Switzerland. The creative roots of a new world with over a hundred vegan organic foods are found here. The LifevisionLab is housed here as well.

The true cost of food

It is expected that in terms of their true costs, organic foods will be less expensive than nonorganic foods.

For many consumers an important incentive to buy nonorganic foods is their cheaper price. Yet today it is a fact that conventional agriculture causes much damage to our health and to the environment, and these so-called external costs are paid by the consumers and the community.

This is not the place to present an overview of all the external costs of nonorganic farming, but one can imagine the extra expenses nonorganic farming generates compared to organic farming, from water pollution to toxic residues in the entire food chain, to antibiotic resistance, soil erosion, soil nutrient loss, desertification and massive climate impacts.

Currently, the discussion primarily focuses on the **external environmental costs**, since the calculation of costs arising purely from health risks is considered difficult. However, costs from health risks certainly do exist, as the daily consumption of pesticide and antibiotic residues from nonorganic foods over many years definitely will not fail to leave its traces on the detoxifying and excretory organs of the consumer, to name but one example.

From the understanding that nonorganic foods are increasingly weakened in their order force (life design principle), it follows that the natural life design principle is able to unfold its life and health-supporting powers only through organic foods. It can also be argued that nonorganic foods should be assigned **external costs arising from their health weakening effects**. Here an important research area is awaiting studies that can supply numbers and arguments for the consumption of organic food, including tremendous savings on health insurance and premium costs.

Given the external costs of conventional farming, the cheap prices of nonorganic foods are an illusion, leading consumers astray in their shopping decisions, both to their own detriment and to the detriment of the community. **According to the 'polluter pays' principle it makes sense to require payment of the full external costs of nonorganic farming.**

Under the terms of the 'polluter pays' principle, the artificial fertilizers and chemical additives used in conventional agriculture should be taxed in order to compensate the community for the costs arising from the ecological and health damage from nonorganic agriculture.

Therefore ongoing investigations by impartial institutions are required in order to determine the private and public costs of nonorganic farming. The organization most suited to tackle this responsible task seems to be the FIBL (the Research Institute of Organic Agriculture, with branches in Switzerland, Austria and Germany), since it possesses the necessary skill and viewpoint for applying both methods.

Organic muesli with organic oat drink and a vegan alternative to organic sour cream (Soyananda)

Whether, when and to what extent these private and public costs can be transferred to the polluter depend on the political awareness of the people and their ability to assert their requirements.

When enough people become aware that the external costs of nonorganic farming incurred by individuals and communities are harmful to both, and furthermore, that the proper allocation of goods is distorted by the low prices of food products from chemical-industrial farming, then the demand for the 'polluter pays' principle will increase.

When true cost-pricing is achieved, it is to be expected that organic foods will have lower prices than nonorganic foods. One scientific study found that if the country of Denmark made a complete shift to organic farming, it would save 134-202 million euros per year in external costs. When the positive external effects of organic farming and the negative external effects of nonorganic agriculture are integrated into the market mechanism and reflected in the product price, conventional foods will be more expensive than organic foods.*)

Organic foods will then benefit from a price advantage. The price alone will increase the demand for organic foods compared to nonorganic foods. Therefore, in the interest of public health and health-care cost reductions it is very important that citizens be informed about these issues and that true cost-pricing for nonorganic foods is pursued.

*) LEBENSMINISTERIUM (2009): Grüner Bericht. Eigenverlag, Vienna. ALROE, H.F. & E.S. KRISTENSEN (2001): Researching alternative, sustainable agricultural system. A modelling approach by examples from Denmark. In: MATTHIES, MALCHOW, H. & J. KRIZ (Hrsg.): Integrative Systems Approaches to Natural and Social Dynamics. Springer Verlag, Berlin.

Instead of closing remarks...
...opening a dialogue

I am overjoyed that this book has succeeded in opening up a perspective onto an invisible aspect of the reality of food, which so far has hardly been perceived. Do you like the view into this new world? It allows us to have a much deeper feeling for our foods, and even for reality itself. What consequences does this have for you?

I hope that professionals, as well as many consumers, will discuss the issue of food quality in greater depth so that a new, holistic quality-awareness can be developed. It would be a dream and a sign of amazing progress if a new understanding of people and foods as one whole, living reality could be accepted on a large scale.

The question about the effects of food quality on personal health is valid. It will inspire interesting projects on quality research, which will be able to offer further insights and greater improvements.

- If it were feasible, would you welcome the idea of having pictures of the order force on foods in order to make informed shopping decisions?
- Would you want testing organizations to take into account also the order force of foods during their research?
- Do you think it possible or even likely that the universal language of crystallized images from the LifevisionLab for the identification of food quality will become popular? Personally I think that initially, innovative organic producers will use it to inspire their customers, but that it will take a long time for official bodies to recognize this method and work with it. The private sector will lead the way, and the speed of progress will depend largely on individuals.
- After reading this book, will you value higher organic food quality more than before?
- In the course of these developments we all wish for food producers in general to be more quality conscious and to start better fulfilling the demands of consumers.
- Like Bhutan, which has decided to allow only organic agriculture in its country, there will be other countries that will make intelligent decisions by cutting themselves off from private interests and feeling a responsibility for the shared common good of good health for humans, animals and the planet. May this deep understanding take root inside many influential people and bring about a global trend that benefits everyone.
- We would like to invite you to take the initiative by informing the public and the local press with an exhibit about the theme of food quality, perhaps in cooperation with a health food store.

Posters for this exhibit can be found on the following pages.

Thank you

It is our hope that people from all corners of the earth will become more conscious of the necessity of feeding ourselves with naturally grown organic foods full of life-energy. We also believe that access to naturally grown organic foods is a birthright that everyone can and should claim for themselves and their families.

We therefore thank all those who have travelled this road themselves and who joyfully contribute their insights from their particular fields in order for this basic human right to be realized step by step and finally for all people. These forward steps will bring about great benefits for the planet and all of its life forms.

We are grateful for the understanding of the significant wisdom of unfettered personal growth and for every single step towards this goal, no matter how tiny.

We thank all organic farmers and gardeners for producing organic foods. We especially thank the pioneers of all the various types of organic agriculture, who developed, explored and taught organic agriculture in spite of all the difficulties they faced, as well as those who are doing this today. We also thank those who acknowledge, value, support and promote organic agriculture in all its forms.

And we thank the Creator for the amazing beauty and the infinite wellbeing in His creation.

We wish you and all people infinite health.
Freely paraphrased from the health definition of the World Health Organization of the UN:

» *We wish for all people to have complete physical, mental and social wellbeing on all levels of reality!*

Inspiration...

Don't blame the world.
Better it.
How?
By bettering yourself at every
moment.

To serve
And never be tired,
Is love.

Only an explosion
of self-giving activities
from each individual
can save this precarious world
from threatening destruction-night.

Each human being is expected to be
a soulful ambassador of goodwill
to improve
the standard of mankind.

Aphorisms and Artwork by Sri Chinmoy

SRI CHINMOY with A. W. Dänzer in front of the entrance to the Soyana company on September 11th 2002. The author is holding a gift from his teacher, a song that Sri Chinmoy composed about the soul of Soyana: "**The Secrets of Perfection-Flames to strengthen and enlighten our earthly frames.**"

Infinite gratitude

to my spiritual friend and teacher Sri Chinmoy,
who has brought forward my soul,
who fills my heart with the presence of the Creator,
who illumines my mind,
develops my life for the service of the whole,
and offers the fruits of my life to mankind.

A. W. Dänzer, Schlieren 2014

About the author

Walter Dänzer was born in 1947 as the youngest of three children in a workers family in a village near Zürich, Switzerland and enjoyed a happy childhood. His father rode his bicycle to work at the food-producer Maggi and served as a local council member for many years. Dänzer studied law and economics at the University of Zurich. After graduating he wrote his thesis as a research assistant in the field of agricultural policy and founded an interdisciplinary working group comprised of economics, agronomy, sociology and psychology, in order to design a reorientation of the outdated agricultural policy, away from chemical and industrial agriculture. He visited many farmers and learned about their problems. In the thesis he wrote in 1974 he proposed a modern orientation for agricultural policy towards health and ecological goals by supporting organic farming, including an annual prize-giving to three organic farmers with a film presented on Swiss national TV. This would generate a large demand for organic foods, which would cause many farmers to switch to organic farming. Unfortunately the thesis was flatly considered unscientific and the interdisciplinary insights were completely stifled. Disappointed, he withdrew from society and lived in the mountains for two years.

In the mountains Dänzer joined a group that was learning to be self-sufficient. The experience of nature with plenty of exercise in the open air brought him from his head into his heart. They all learned the necessity of working on themselves. That winter in Zürich he discovered his spiritual path and went back to the mountains for another year, this time all by himself at an altitude of 1,500 meters. The simple life in nature combined with an intense meditation practice gave him a deep faith in the indestructibility of nature and a practical direction about his role in society.

After three years of working as a sales and advertising manager in the publishing world, he founded the Soyana company in 1981. He was the first in Switzerland to offer plant-based meat alternatives and also introduced soy meat and tofu through two bestseller books and hundreds of cooking classes. Dänzer and his team have by now developed more than 100 vegan organic foods, popular in Swiss health food stores. Soyana's innovative and high quality organic foods are also increasingly in demand in surrounding countries.

Inspired and supported by his friend and teacher Sri Chinmoy, and as part of his holistic spiritual life, he has performed as a singer in hundreds of concerts of soulful music in many countries, inspired seekers to lead a spiritual life through free lectures and classes and ran his fastest marathon as an athlete at age 41 in 2:31, earning a top 10 place in a field of 5,000 runners.

In January 2013 he opened in Zürich Switzerland's first vegan and organic restaurant. Among other things the restaurant also offers Vegelato, a varied assortment of top Swiss quality vegan organic ice cream.

Soyana: the search for "peace food"

"The Secrets of Perfection-Flames", Soyana is a small family business that produces food out of an inner aspiration towards perfection and a loving harmony with the earth and all its inhabitants.

Founded in 1981, it was the first company in Switzerland that offered a meat alternative. The Soyana concept—making it easier for people to nourish themselves in a healthy way through plant-based foods in harmony with nature—became a success. Soyana foods supported thousands of consumers to make big positive changes in their lives.

After starting with organic tofu and tofu burgers, we developed other innovations. Today we offer over 100 vegan organic foods as healthy and tasty alternatives to meat and dairy products. Among them are DINKI, an excellent meat alternative made from spelt, the SOYANANDA line with vegan cream cheese, sour cream and yoghurt, the SWISS DRINKS, a wide variety of soy, rice and cereal drinks, and of course our CHI, the vibrant, fermented organic drink in four flavors and one of our top products. Recently we have also added VEGELATO, 50 different kinds of vegan organic ice-cream sold in our VEGELATERIA!

Please visit us at *www.soyana.ch* and *www.vegelateria.ch*

Supported by the research in our LifeVisionLab we are working hard to continuously improve the unique quality of our Soyana foods. We hope that our discoveries will contribute to making high quality the norm in the food industry.

4 important principles of Soyana foods

Organic quality // All Soyana foods are produced only with ingredients from certified organic agriculture, free from genetic modification.

Plant-based ingredients // Plant-based foods are healthier, more economic, loving and sustainable. Through plant-based foods our planet can feed so many more people than through animal foods.

Energetic quality // The Ojas Harmony System installed in our production facilities ensures that the excellent organic quality is not compromised by technological influences.

Good consciousness // We hope that the inner balance and joy of our workers, based on their healthy lifestyle, prayer and meditation, is transferred to our foods to strengthen and enrich our customers.

On the next two pages you will find some examples of Soyana foods. Healthy food should also be a delightful experience!

Chi
Chi is organic
fermented
unpasteurized
made with love
Roots
Chi
Fruits
It's Chi Time
the life force reveals many flowers-of-life
you are drinking pure life-energy!
Chi
Flowers
Chi
Green tea/Herbs
bio
1000x © LifevisionLab
1000x © LifevisionLab
1000x © LifevisionLab
1000x © LifevisionLab
Cola drink
1000x © LifevisionLab
Chi - toasting with the finest raw food elixir
soyana
Chi lives
Chi gives you amazing life-energy
Fruits
Green Tea/ Herbs
Roots
Olivier Bernhard
World and Swiss Champion
IRONMAN
SWITZERLAND
Bernhard Olivi 8:16
Chi stimulates!
Fruits
Green Tea/ Herbs
Roots
"I use Chi primarily after training. Because of its high quality Chi is easily digestible, even during exercise. In endurance sports like ultra running and marathon swimming Chi has proven to be more gentle to the stomach than normal sports drinks.
Even when driving home after a long working day Chi keeps me awake, because it's so stimulating and refreshing."
"Chi shortens my recovery time after an intense workout or race.".
www.soyana.ch

POSTERS

Posters for inspiration and exhibits

A SERIES (14 THEMES)

A-1

A-2

A-3

A-4

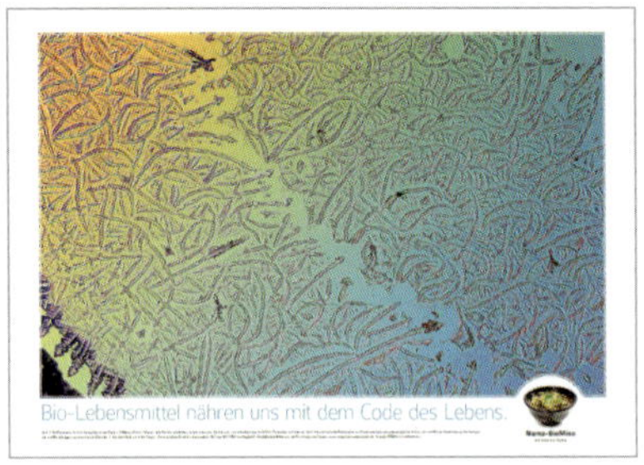

A-8

A-9

A-10

A-11

B SERIES (14 THEMES)

B-1

B-2

B-3

B-4

B-8

B-9

B-10

B-11

The most beautiful LifevisionLab images of organic foods from this book are available as posters, sized 20" × 28" (50 × 70 cm). Please note that the titles under the corresponding image of the 28 posters are in German. However, the free exhibition poster for announcing the exhibition and the two

*) Get 7 posters for free when ordering 14 or 28 posters for your exhibit:
5 copies of the exhibition-poster (to poster for your exhibit) and 2 introductory posters with text in English.

A-5

A-6

A-7

A-12

A-13

A-14

B-5

B-6

B-7

B-12

B-13

B-14

free introductory posters are printed in English. Each order will be supplied with a separate sheet with the English translation of all 28 posters' text.

To order, please visit
www.organic-nonorganic.info